CORAL REEFS

Coral Reefs

Majestic Realms under the Sea

PETER F. SALE

Yale
UNIVERSITY PRESS
NEW HAVEN AND LONDON

Published with assistance from the foundation established in memory of
Amasa Stone Mather of the Class of 1907, Yale College.

Yale University Press books may be purchased in quantity for educational, business,
or promotional use. For information, please e-mail sales.press@yale.edu (U.S. office)
or sales@yaleup.co.uk (U.K. office).

Set in Electra type by Integrated Publishing Solutions.
Printed in the United States of America.

Library of Congress Control Number: 2020947101
ISBN 978-0-300-25383-2 (hardcover : alk. paper)

A catalogue record for this book is available from the British Library.

This paper meets the requirements of ANSI/NISO Z39.48-1992
(Permanence of Paper).

10 9 8 7 6 5 4 3 2 1

For my mentors, my colleagues, and all those students who helped me learn to appreciate coral reefs;

For my family, Donna, Darian, Michelle, and especially Isabelle, who will get to see how the Anthropocene pans out;

And most of all for the coral reefs themselves, which gave me a place and a reason to be.

CONTENTS

PREFACE

In April 1999, I was in Fort Lauderdale, Florida, attending a scientific conference hosted by the newly formed National Coral Reef Institute at Nova Southeastern University. About five hundred coral reef scientists were gathered to explore the science of assessing, monitoring, and restoring coral reefs. There were the usual plenary talks, contributed papers, and lively poster sessions, where enthusiastic students and professors display posters describing their current research and try to entice other participants with their elevator pitches: "My work is really exciting because . . . ."

Poster sessions are great times for conversation, about the posters and about a host of other topics. The coral reef science community is a widely scattered but closely knit community of colleagues, collaborators, and friends. In the best conferences (and this was one) the conversations are helped along by a liberal supply of alcohol and snacks. I don't remember my paper, or much else about that conference, but I do remember those conversations, because so many of them centered on the circumtropical bleaching event that had just ended late the year before.

A mass bleaching event is a rapid and conspicuous change on a reef, caused by physiological stress being experienced by the corals. The stress leads to a breakdown of the most important symbiotic relationship on reefs, that between the corals and their microscopic algal symbionts. The algal cells (hundreds of thousands to millions per square centimeter of coral tissue) are pigmented, and as they are expelled the coral turns a ghostly white.

This can happen overnight. In 1997–1998, during the then strongest El Niño event on record, coral reefs had turned white around the world. It was the first time bleaching had occurred on such a massive scale. Over the weeks after bleaching, many of those corals had died, and there were reports of reefs in all oceans that had suffered extremely high die-offs of coral. The Fort Lauderdale conference was the first opportunity many of us had to talk with colleagues who had witnessed this destruction. We all understood the seriousness of what had happened, and I remember that we all anticipated that a signal as conspicuous as what had just occurred would surely be a sign to the world that we had to deal with climate change, and quickly. We found that expectation comforting. Except that it did not happen. Most of the world just did not care.

This situation is deeply unfortunate, because as well as being ecologically and economically valuable ecosystems, coral reefs are a supremely magnificent feature of our planet, majestic oceanic realms with amazing stories to tell. Their stories can bring pleasure, open us to new perspectives, and delight our minds and senses while revealing intricacies in the way this world we inhabit is assembled. Reefs offer stories of sublime beauty and of great power, as well as stories that can help us understand our current predicament as we alter our world in ways that will do harm to us and those we love. Remembering the conversations in Fort Lauderdale, I knew I needed to write, because I believe we need those stories. This book is about coral reefs and what they say to me. It is also about what is happening to reefs, about why, and about what this says about us. But it is mainly about what reefs are and how we need to appreciate them.

Why need and appreciate? I say "need" because I am convinced that humanity must relearn something most of us have long forgotten—that we are a part of the biosphere on this planet, not external to it. Connecting with natural ecosystems can help us do that, and reefs offer many ways in which we can connect. I say "appreciate" because, despite being regularly deluged with stunning images, evocative videos, and detailed descriptions in the media and all over the web, we remain largely ignorant of what reefs are or why they are worth caring about at all. Reefs are part of the Other, not part of our world, and one brief snorkel tour during a beach vacation doesn't build an emotional connection. My goal is to get at least some few more of us to know reefs a whole lot better. Only by knowing them better will we truly understand the important messages they bring. Understand, and care about, those reefs and their messages. Appreciate them.

Despite having been around for a very long time, coral reefs are perhaps the most sensitive of all ecological systems, easily disrupted or destroyed by changing circumstances or human action. Coral reefs are being very substantially altered by the stresses we are causing today. In 2011, I detailed this story in *Our Dying Planet*, stating that by 2050 coral reefs as I knew them in the late 1960s will simply not be present anywhere on this planet.

Think what that statement means. I wasn't saying that *some* coral reefs will be degraded. I was saying that *all* coral reefs could be lost. It's like saying that *all* rain forests around the world will soon be pastures. We have never yet had that profound a global impact on any major ecosystem, yet we could eliminate coral reefs in just three more decades! I still fear that. There remains a vanishingly small window of opportunity for us to alter our behavior, reduce the impacts we are having on the planet, and help a few lucky reefs to persist and maybe to flourish, but we show little sign of wanting to take advantage of that window.

I was not the first reef scientist to warn of the loss of coral reefs. One of the plenaries at that 1999 conference was titled, "Is It Time to Give Up?" It was subsequently published (R. W. Buddemeier, *Bulletin of Marine Science* 69 [2001]: 317–326). In it, Bob Buddemeier of the University of Kansas argued that we needed to practice triage to prioritize where to put our science and conservation efforts because we had to face the fact that most reefs were now toast. Several of us have done our best since to try and articulate just what we humans are doing and the likely consequences. Our efforts have failed.

Since 2011, I've reached the conclusion that patiently explaining the cost of continuing our present patterns of behavior is just not working to get us to move. Like deer in the headlights, we stand, and we wait, and we watch, and then we take token steps or continue as before.

And so, I turn to stories from a coral reef. Because I believe that, if we can reconnect with the natural world, then maybe we can come to actually appreciate, in a visceral way, that our civilization cannot survive on this glorious rocky planet hurtling through the universe if we go on as we have been. We must take care to preserve the capacity of ecological systems to sustain their integrity and resilience.

The biosphere is not simply here, one fact about our world; the biosphere makes it possible for us to be here, and reefs are sublime portions of the biosphere. In fact, reefs are so sublime, fascinating, and marvelous they make connection easy. But first we need the stories. I believe that the wonder that reefs can inspire can lift our spirits and drive the reforms that we must make. Just possibly, reefs might help us act sufficiently fast to save them too.

♦♦♦

Apart from a brief account of my own awakening, this book proceeds from accounts of coral reefs as natural phenomena, to aspects of how reef ecosystems function, and to how we perceive and value reefs. The final chapters deal with the important message that reefs have been trying to send, why we have failed to notice, and what we need to do to right the damage we have been doing to the biosphere. I am hopeful that we will be inspired to act forcefully and soon. My stories are mostly ones of how our comprehension of coral reefs has grown with the growth in reef science. Many of the stories also have a chronology driven by the way in which scientists pursued them over many years, and some of the later tales are more recent in the sense that the underlying science began to be explored more recently. In telling these stories, I have tried to tell of the discovering as well as the discoveries, because I find accounts of how science is done far more compelling than an enumeration of the results. Some of the science discussed is my own but most is not, and the focus on scientific understanding is simply a consequence of how I have always approached the natural world. This book is my effort to reveal the wonder that coral reefs can inspire. Come with me on this journey!

CORAL REEFS

ONE

Beginnings

It's 1998. I'm on a floating island south of Cancún, Mexico, witnessing the clash of mass tourism with a coral reef. The plastic palm tree near the snack bar hints that the island is not real. The masquerade creates a platform from which, over the years, thousands of tourists have made their first acquaintance with a coral reef. The quasi-island offers change rooms, showers, toilets, a real bar as well as the snack bar, plus plenty of space to stand about and talk. A never-ending wagon train of Sea-Doos, kayaks, and other craft plus an ancient ferry deliver tourists and collect them again to whisk them back to their hotels; supplies arrive, and effluent is discreetly removed by separate service vessels. Waves of novices don snorkeling gear and stumble down steps into the ocean, there to bob about, peering at the coral and the fish, periodically bumping into or standing on the reef, while trying not to drown. They emerge bruised, battered, sometimes water-logged, always elated. The reef suffers quietly.

The island swarms with excited people of all shapes, sizes, and ages, in various stages of undress, eating, drinking, applying suntan lotion, preparing to snorkle, or recovering from their excursions. And talking incessantly, as people do when not glued to their devices.

"Did you see that big blue fish?" a twenty-something, slightly wasted blond dude asks, while checking his reflection in a nearby window.

"With yellow blotches? No, . . . don't think so; big googly eyes?" his seriously sunburned girlfriend replies. "I guess I thought there'd be more color; couldn't tell what was coral and what was just rock. Everything seemed slimy."

"You touched stuff? I was afraid to put my hands out anywhere near anything; it was all so weird," chimes in their friend, whose white bikini nicely sets off her tan.

"Something scratched me. Look," offers blond dude. "Guess we should be heading back to the hotel for happy hour."

They drift off out of my hearing. I wonder what they'll remember.

Most of us have never had the luck to see a coral reef, and most of those who have are just like those tourists. They've seen a reef once during a quick snorkeling opportunity wedged into a busy sand-and-sun vacation somewhere far from home. We all think we know what reefs are. We've seen countless images and videos of them, streamed across our media. But we don't really comprehend reefs, don't understand them, and they haven't captured our emotions. Because of this they remain part of the Other. People do not care about the Other, and that is most unfortunate.

We are now far past the time to comprehend what we are doing to coral reefs around the world, and what we are very close to losing. We are rapidly removing coral reefs from the planet, yet most of us remain barely aware. The powerful messages coral reefs have been sending us about the state of the planet simply have not jolted us into action the way they should have. Our wholesale destruction does not just matter for coral reefs; it matters for the wider biosphere (including us), for how we act in future years, and for our own long-term survival. If you find this a bit over the top, keep reading.

A coral reef is fundamentally improbable. By any reasonable measure, coral reefs should not exist, and yet they do. Earth's oceans have been enriched by flourishing coral reefs over most of the past half-billion years; by their existence, reefs testify to the resilient fecundity of life, its proliferating diversity, and its seemingly limitless capacity to endure even as the world changes. Particularly now, when the full extent of our human-caused global environmental crisis is becoming clear, the long-term success of coral reefs should inspire us, providing the strength and determination to mend our ways and steer our planet toward a better future. And yet that success does not inspire. Because we do not know reefs. We only know *of* them, and most of us scarcely know that.

♦♦♦

I think I was preadapted to become a coral reef ecologist. I was born in and spent my earliest years in the tropics. I loved the ocean and warm sunshine. I loved watching animals and wondering what they were doing and why. I don't know where my love of nature came from, although children did spend a lot more time outdoors back then, and with plenty of idle time to let their imaginations wander. After a gritty couple of years in London in the final years of World War II, a time I remember chiefly in shades of gray, our family of four returned to the tropics, or almost tropics, arriving in Bermuda late in 1946, where we lived until early 1951. In Bermuda, my learning about coral reefs began in earnest.

I did not know at the time that I was embarking on a lifetime career in coral reef science. I left Bermuda at the age of ten, but I actually learned a few important things during those five years.

My family rented one wing of the main house on a small dairy farm on the south shore of Smith's Parish, and as children did back then, I spent most of my time when not at school playing outside. Some of that play was simply hanging out at the cow barn, helping to milk the cows, distributing feed, and mucking out stalls, but much of it was an exploration of my amazing subtropical world. I climbed trees, explored abandoned buildings, and journeyed through the less-used part of the farm, all while immersed in a richly imagined amalgam of Tarzan, Roy Rogers, and Captain Hook. Whenever an adult could be dragged along, I spent time at the small beach at the base of the low cliff just a cow pasture away. Google tells me I could now stay in half of my house for a Bermuda vacation at twelve hundred dollars a day, but a housing development has replaced the farm, so the sense of my own private beach is probably gone. (Other than a fresh coat of paint, the house appears to have changed little over the years, although I hope they have upgraded the plumbing.)

Our beach was tiny, and it usually had quite strong wave action (as judged by a skinny kid standing barely more than a meter tall), more a case of scattered patches of sand among the rocks. This beach was definitely not one of

Figure 1. The budding scientist at one of our tidal pools in Bermuda. My clothes suggest that it must have been a cool afternoon following school. Photo © Peter Sale.

those sweeping curves of pink that built Bermuda's tourist economy. It had some extensive rocky benches, with tidal pools, and it is in those pools that I began to discover coral reefs (fig. 1).

You see, Dad liked to fish. He was not particularly good at it, but he used to alternate between swimming with his mask and spear (no snorkels back then) and standing on the rocks using a hand line baited with snails that he had bashed out of their shells. My brother and I got to stay well back from the breaking waves, with our own hand lines and the tiniest possible hooks, also baited with bits of snail. We fished the tidepools, occasionally capturing sand gobies that we then struggled to free from the hooks and release, hopefully to survive. (Truthfully, my brother was four years younger than me, so mostly he just got in my way, although he does not remember it that way.) Fishing for gobies can wear a bit thin after a while, but exploring what was in those tidepools . . . that never got boring. I learned early that if you were quiet and watched carefully, there were lots of critters busily going about their business there. Best of all, with a face mask you could get right into the pool and really see what the animals were doing.

The bleeding tooth snails were abundant on the rocks and were my preferred choice as bait. *Nerita peloronta* is named the bleeding tooth because the inner edge of its aperture contains two toothlike ridges, white in color and surrounded by a rather anemic orange shade. I knew that the teeth were not really bleeding, but I still wondered why the snail had those teeth, what it did with them, and why the color around them was vaguely bloodlike. I also wondered why the snails would withdraw their eyestalks if you touched them, and how they could do this, and what it must feel like to have your eyes stuck out on the end of long stalks yet be able to pull them deep inside your body. Never mind pulling your whole body inside a shell and slamming the operculum shut like a trapdoor to keep little boys' fingers out. I also wondered how they crawled over the rocks and what they were eating. I did not kill them indiscriminately, but I knew they made good bait, even if Dad was so unsuccessful as a fisherman.

The rocks around the tidepools were also occupied by limpets and chitons. Somehow, I knew that these were distantly related to the bleeding tooth and much more distantly related to the gobies. But limpets and chitons were nearly impossible for me to dislodge from the rock, and they did not seem to do much either. (I did not know then that they moved about during high tide, feeding over the rock, and then usually returned to the exact same sites to rest at low tide, clamped down tight to the rock keeping moisture inside despite the warm sun.) The bleeding tooths, by contrast, could be active whenever submerged. This made them more interesting creatures, and my interest in animals that visibly behave stayed with me into adulthood.

There were lots of other creatures in the tidepools, but I did not know much about most of them. Occasional small corals lived in the deeper pools, as did sponges, coralline algae, and other sessile creatures that I realized were alive. I knew little more than that, and these creatures seemed so foreign that I could not connect with them. Starfish moved slowly and brittle stars more swiftly, but their radial design left me confused. Where was the head? Could they see? Small crabs, by contrast, or the beautiful red and white banded coral shrimp (*Stenopus hispidus*) that I found one day, all had arms and legs, eyes, and a mouth and seemed to do animal things. The banded coral shrimp was festively colored and waved her claws and white antennae at me, signifying what, I did not know. I certainly did not know that she would also wave at passing fish, which would then pause, allowing her to clamber on and pick parasites off their surfaces, even between their teeth. If I'd known that, I would have wondered how the fishes knew to pause. Indeed, I still wonder that.

Occasionally, I'd come across the newly shed shell of a large Sally Lightfoot crab (*Grapsus grapsus*), and somebody, perhaps my parents, told me how crabs had to shed their shells periodically in order to grow bigger. I tried to imagine what it would feel like to shed your skeleton and be all soft and floppy—I figured it made more sense to have your skeleton inside, but I also thought it might be fun to have all that armor on the outside.

Best of all were the fishes. The tidepools always had sand gobies in them, but occasionally they would contain more brilliantly colored fish as well. Fish like the yellow-banded sergeant major (*Abudefduf saxatilis*), which paraded in shimmering regiments in the water just beyond the edge of the rock shelf—water that I was allowed to swim in, with my face mask, on calm days. Still, the sand gobies were always in the pools, and I interacted with them to a greater degree because they ate our bait. They were probably *Bathygobius soporator*, but several other species occur there, and at seven or eight years old, I was not into counting fin rays or gill rakers or checking to see whether the animal's tongue was notched in order to figure out which species of sand goby I was watching. The gobies spent their days moving about slowly over the sand, occasionally taking a mouthful of sand and letting it pass out through their gills. I guessed that they were eating something in the sand. Many years later, an Australian graduate student of mine showed that similar gobies winnowed mouthfuls of sand, allowing the large sand grains to fall out past their gills while somehow keeping and swallowing the much smaller crustaceans that lived among the sand—sort of like taking a mouthful of tiny pebbles mixed with even smaller bits of candy, swirling them around with your tongue, and then swallowing the candy while spitting out the pebbles. Even if I had gill slits, I think I'd find that a demanding task.

Those tidepools gave me an opportunity to see some of the creatures that lived on reefs up close and taught me something about the different types of animals. The surrounding rock sometimes showed its coral origins, too, retaining the skeletal features shared with the living corals nearby, and this taught me other lessons. I came to understand that all rock in Bermuda was of biological origin. I don't know whether I learned this in school or elsewhere, but I knew that Bermuda existed as a reefal structure, built by corals, that had grown around the peak of a volcano rising from the seafloor. I understood that where there was evidence of coral in the rock above high tide level, this was because of uplift in past ages. I did not know the details or the complexity of Bermuda's geology, and some of what I thought I knew as a child was not correct, but I realized that I lived on a rather special island.

♦♦♦

When I could not be at the beach, an abundance of other animals (lizards, caterpillars, wasps, birds) kept me amused. One gave me a lesson I did not understand until many years later. The Bermuda land crab or blackback land crab (*Gecarcinus lateralis*) occurs in Bermuda, Florida, and a wide swath of the Caribbean. Unlike most crabs, this is a sturdy terrestrial soul that digs burrows up to a meter long in sandy soils. Deep within the burrow, there is sufficient moisture for it to keep its gills moist. It comes out to forage at night, primarily on plants.

To me, this was a crab that lived in a burrow and might come out if it was teased in the right way. The right way required finding a stem of grass, preferably a long one topped by a seed head. By crouching down, being very quiet, and poking the grass stem into the burrow, I could sometimes get the crab inside to swat at it, to grab it with one of its claws, and maybe to take a few steps toward the entrance. Rarely, I could get it to reach the burrow mouth, but I don't think I ever got it to come outside. My friends and I would spend what now seems like hours attempting to entice land crabs out of their burrows. What we planned to do if a crab really came right out, I have no idea. Why this was so entertaining fascinates me now. Of course, we did not have cell phones or even television to keep us entertained, so I doubt that many kids engage in this fruitless pursuit today. Fruitless it may have been, but it allowed us to explore and to think about other creatures that shared our world. Why did they dig burrows? What did they do all day? Why did they not live down at the beach?

The big lesson came one evening after dinner when, contrary to our usual pattern, my parents took my brother and me down toward the shore. I think it was because a hurricane was coming, the seas were up, and it would be fun to watch the surf. I know now that this particular evening occurred during one of the summer spring tides (the highest tides of the month), but back then I did not even know there was such a thing as a spring tide.

Now, the Bermuda land crab lives well away from the sea and is quite common. There were land crab burrows in our lawn and all through the pastures. This evening, it seemed as if every crab on the farm was out of its burrow and moving down toward the ocean. They had not been out when we went down to the shore, but as we were heading home in the twilight, crabs were everywhere. As our flashlight beam shone on them, they would stand up tall, stretch their two claws wide and high, open them, and wave them menacingly at our ankles (fig. 2). Switch off the flashlight, stand quiet, and they would settle down and recommence their oddly diagonal walk down toward the shore. What were they doing, and why?

The standing up tall and waving of claws is what every crab in creation does when cornered. It looks menacing, looks like it means business. In short, it is a defensive display that crabs have evolved, which does the job it is intended to do. It certainly stopped us. And it stops most creatures smaller than us.[1] In a way, it's part of a widely understood body language used by crustaceans (crabs and their kin) and by all vertebrates, including us—stand tall, look as big as possible, display any weapons you possess—just think *West Side Story*. (Insects, although related to crustaceans, do not do this. They tend to do things like suddenly exposing color patterns that resemble big eyes or are simply suddenly bright.) Many fish spread all their fins and

Figure 2. Standing up tall, stretching its claws wide and high, opening them and waving them menacingly at our ankles . . . . This image shows *Gecarcinus lateralis* in its characteristic threat posture. It was photographed in summer 2000 at the San Miguel Reserve, Puerto Rico. Image © Donald L. Mykles, Colorado State University.

then turn sideways to the intruder to maximize their size. Frogs, lizards, and the hog-nosed snake puff themselves up by taking a couple of extra-deep breaths; so does the puffer fish by swallowing water. Birds fluff their feathers, and you and I raise the hair on the back of our necks—a rather futile enlargement technique, but our more hirsute mammalian relatives use the standing up of hair or fur very effectively in such circumstances.

So, those crabs were threatening us. But why were they walking down to the ocean? I had no answer for many years (and I never again saw them doing this). The Bermuda land crab is one of a relatively small number of crustaceans that have invaded the land, and they are well adapted to their terrestrial life. Their burrowing habit gives them the moist environment that they absolutely must have, because they breathe using feathery gills tucked underneath their carapace. They have almost succeeded in breaking their dependence on the ocean. But not quite: their eggs hatch into aquatic larvae that spend a period in the ocean before reaching the juvenile stage and emerging onto land. The crabs we saw that night were mostly females making their annual migration, walking down to the ocean on a spring tide evening.[2] Once there they would wade in just far enough to wet their bellies, open their tucked-up abdominal segments, and wave their swimmerets around to disperse their eggs, which would promptly hatch into tiny larvae that would swim or be carried away on the outgoing tide. Only after completing their larval life would the few surviving juveniles come back to shore, wander up the hill, dig burrows, and start the cycle again.

Crabs copulate and have internal fertilization. At copulation the male crab delivers sperm to the female; she then holds these sperm within her body for several months until she needs them. Aquatic crabs copulate in the water, usually just after the female has molted. The Bermuda land crab, however, keeps to itself, deep in its burrow, during molting time. Instead, this crab copulates on land, usually at the mouth of a burrow built by a male not far above the high tide level, during the reproductive season. The females, having released their young in the surf, walk back up the beach to be met by amorous males. Several months after copulation, back in her own burrow, each female ovulates, uses the stored sperm to fertilize her eggs, and then extrudes the eggs and fastens them in among her swimmerets—the tiny legs on her abdominal segments, segments that most people do not realize exist because they are kept tucked up tightly under the crab's body. This again is exactly what aquatic crabs do, but the Bermuda land crab does this on land deep in the burrow. Packed in a moist space under the tucked-up abdomen, the eggs develop, but finally the time comes for them to hatch, and the evening migration begins.

That it took me years to learn what was going on that evening makes the event that much more momentous for me. At nine or ten, I thought it was amazing that "all the crabs on the island were walking down to the sea." Now I know it was mostly the female crabs, and I know they were on an important mission.

♦♦♦

As a child, I was intrigued by the natural world, and my parents never discouraged my interest. I sense that most children today are curious about the natural world, but I fear that many of them have few opportunities to give this curiosity free rein. They explore and they learn, but not about the real environment in which we all must live. Coral reefs are full of stories, grand stories of mountain building and sea level rise and fall, as well as smaller stories of simple creatures doing amazing things because that is who they are. Reefs can teach us about time and space and about the amazing richness that is the biosphere. Life is made more wonderful because reefs are here.

My years in Bermuda allowed me to begin to know the seashore and the reef. I learned about some of the animals, but I also began to learn about deep time and complex geological processes. I learned that it was possible to live on an island built of rock manufactured initially by tiny corals. Had I grown up elsewhere, I'd have learned fascinating things about that type of place. In fact, I did just that when I moved from Bermuda to Canada. Then, after thirteen years in a temperate environment far from any ocean, I had grown up, completed bachelor's and master's degrees in zoology, and discovered that one could be paid to explore the natural world! I then traveled a third of the way around the planet from home and back to the tropics to become a coral reef scientist. If my early education had been serendipitous, so was the decision that led me, in mid-August 1964, to board a DC-9 in Vancouver, Canada, bound for Honolulu and doctoral studies at the University of Hawai'i.

To head off to Hawai'i without even a professor identified in whose lab I would work, I had turned down the opportunity to study with outstanding researchers at the University of Miami or at the University of British Columbia, two of the strongest campuses in marine biology at that time. Until the possibility of studying there arose, I did not even know there was a University of Hawai'i! The university offered me a teaching assistantship. Hawai'i offered me Polynesia and coral reefs. I chose with my heart, not my head, and began learning how to pronounce Hawaiian place-names before I even set off for the West Coast and that plane. I don't recommend this as the best way to make decisions concerning graduate school, and yet this is how I began a lifelong career as a coral reef scientist, a career that would take me around the world. One of the craziest yet best decisions I ever made—choosing Hawai'i over UBC or Miami—and choosing a lifetime of discovering the wondrousness of coral reefs.

TWO

Serendipity in Deep Time

My graduate studies in Hawai'i were a wonderful time of learning in and out of the classroom—my first chance to be back in the tropics, armed with the beginnings of an education in the life sciences. I remember field trips to see all those incredible marine invertebrates I previously knew only from books or as colorless, fragile specimens stored in alcohol. In fish biology, I disassembled and then reconnected the skeleton of a small skipjack tuna—now I know how many bones it takes to build a fish—and I took part in Professor Bill Gosline's annual fish-collecting expedition at Waikīkī. Many years previously, state officials had given him a permanent statewide collecting permit for fish in any number using any means! They no longer issued such loosely worded permits, and Bill enjoyed tweaking their noses by taking his graduate class of twenty or so to collect fish, using rotenone poison, right in front of the tourists basking on the beach.[1] It was surreal to be scooping up dozens of meter-long moray eels, plus lots of smaller fishes, from the water where those tourists had recently been wading, although probably not the best form of public relations for the university. At least we did not poison the intake waters for the Waikīkī Aquarium that year. The collected fish got used during the course as we learned to identify the many species, learned the varied arrangements of jaws and teeth, and got to see, in the specimens in front of us, the multitude of ways that fish have solved the problem of building an effective tail. It's an engineering challenge to build a strong propulsive structure on the end of a tapering, flexible column of vertebrae. That we collected more than a hundred different species from one small section of reef impressed this Canadian used to far fewer kinds of fish in a lake or river.

My learning did not stop at biology. Hawai'i was also Polynesia, and this was before the invention of Boeing's 747 delivered mass tourism. Even O'ahu had its remote places—secluded beaches, deep, silent valleys, small settlements where Hawaiian was still spoken, and high mountains where native birds and native tree snails could still be found. The Hawaiian Islands share with Bermuda a volcanic origin, but the volcanos that made Hawai'i are much younger and still stand proud; the land above sea level is mainly volcanic basalts, and underwater sites are shaped far more by lava flows than by reef building. For me, it was like arriving on a new planet. I reveled in the sights, sounds, tastes, and smells, and sought to learn as much as I could about the reefs, the islands, the people, and the culture of this special place. That my time there coincided with *Sgt. Pepper* and the Summer of Love simply added to the sense that everything was exciting and new.

Delving into Hawaiian culture led me to Polynesian history, which led me to Captain James Cook. I found time to read about his voyages and wonder at what he found scattered across the Pacific. A couple of years later, I stumbled on Cook's published journals: three heavy volumes compiled by New Zealander J. C. Beaglehole, covering all three of Cook's circumtropical voyages. Doesn't sound much like psychedelia, and don't ask me why, but I read them over the next few weeks, even the pages of entries, day after day, of "Winds freshening from the south at first light. Fed slops to the men." I was fascinated at how Cook was able to understand and respect the peoples of Tahiti and Hawai'i, and even the warlike New Zealanders, but had great difficulty seeing the humanity in Aboriginal Australians. While Tahitian and Hawaiian societies were organized hierarchically with nobles, priests, and commoners, permanent settlements, and rich material cultures, the Aboriginal Australians had little in the way of material culture, permanent settlements, or clearly evident societal structure. Cook and the Indigenous Australians had no common frame of reference. It is difficult to appreciate the Other—also a part of our problem with coral reefs.

Cook's journals also revealed what he thought of coral reefs. Perhaps not surprisingly, he had little good to say about them, but even Charles Darwin had trouble understanding coral reefs when he came upon them fifty years later. Imagine the difficulties Cook faced off the northeast coast of Australia, far from home, totally dependent on the wind to propel his ship, sailing uncharted waters where reefs, ready to rip the belly from his vessel, might lurk just below the surface in otherwise deep water. For someone in Cook's position, reefs can be frightening, threatening, unanticipated agents of utter disaster, and Cook came very close to losing his ship. He had no time to marvel at the fact that reefs exist.[2] Fortunately, we are not in Cook's position, and we can take time to marvel.

♦♦♦

It's tempting to think of coral as architects and coral reefs as the architecture—all those hard-working coral polyps toiling together, year after year over millennia, to build magnificent marine cathedrals. That is far too simplistic. For one thing, the corals are not building reefs: they are simply constructing their own skeletons as they grow. The reefs are the accumulation of all those skeletons. But even that is too simple, because many creatures besides corals contribute to reef building, and all reef building includes a healthy portion of destruction. The true story is more complicated, much more haphazard, and one that makes the fact that reefs exist far more amazing.

I've come to believe that to really appreciate coral reefs, it is necessary to appreciate both deep time—expanses of millions of years—and the power of the repeated operation of small-scale, short-term processes to construct imposing structures. In this sense, a coral reef is an erratic process rather than a thing: a process that extends over immense periods of time through the action of a multiplicity of forces, some of which build while others destroy and most of which are tied to the lives of tiny individual organisms. A reef is serendipity in deep time. If that seems far too metaphysical, let me start at the beginning with some basic biology and then move on to geology and history.

Among the multitude of creatures on this planet is a very old group called the phylum Cnidaria (with a silent C reminiscent of the Gnu's silent G). These corals, anemones, jellyfishes, and their relatives share a unique feature: they all possess nematocysts scattered over their surfaces. The nematocyst is a tiny, diabolical weapon: a toxin-bearing, spearlike, sometimes barbed projectile on a long, coiled filament that is housed within a specialized cell, the cnidocyte (hence Cnidaria). When stimulated by touch or other disturbance, the cnidocyte ruptures, releasing the spear under pressure that propels it toward the source of stimulation. That's why many corals feel sticky to your touch—you've been harpooned by the nematocysts. No other type of creature has anything resembling the nematocyst. Ancient cnidarians invented the nematocyst half a billion years ago, and they have used it ever since to entangle and pacify food organisms, ward off predators, and fight with neighbors. In some cnidarians, especially the box jellies (class Cubozoa), the nematocysts are sufficiently toxic that a grown human can be killed by the stings of multiple nematocysts if entangled in the long, trailing tentacles. If you see or are stung by a creature with nematocysts, it's a cnidarian.[3] Their ancestors can be traced back at least to the Cambrian period, and some specialists believe that they first evolved in the late Precambrian, more than 570 million years ago.

Modern cnidarians are divided into four classes of creatures imaginatively named the flower animals (Anthozoa), water animals (Hydrozoa), drinking cup animals (Scyphozoa), and boxlike animals (Cubozoa). Two of these groups, Scyphozoa and Cubozoa, are only rarely colonial and spend most of their lives as what we'd call jellyfishes living in the water column. The other two spend their lives as polyps attached to the substratum; these are usually colonial, but even though what we call corals overwhelmingly belong to the Anthozoa, lots of noncoral Anthozoa exist and a few corals belong to the Hydrozoa.[4] Life is not meant to be tidy!

Nearly all modern reef-building, or stony, corals belong to the order Scleractinia, a large group of cnidarians characterized by the sixfold radial symmetry of its polyps. If you peer into the mouth of a polyp of the Caribbean elkhorn coral (they have large polyps), or better still, if you look carefully at the skeleton of a dead polyp, you can see that each round cup has a series of six vertical, wall-like septa projecting in toward its center, with six smaller septa between them and twelve more, even smaller, between those. (In the nineteenth and early twentieth centuries, when biologists studied corals by

Table 1. Relationships among the corals and their cnidarian relatives.

Phylum	Class	Subclass	Order
Cnidaria	Anthozoa	Hexacorallia	**Scleractinia** (true stony corals) Actiniaria (the anemones) Zoantharia (the zoanthids) Corallimorpharia (the corallimorphs) Antipatharia (the black corals) **Rugosa** (fossil rugose corals) **Tabulata** (fossil tabulate corals)
		Octocorallia	**Helioporacea** (blue coral) **Alcyonacea** (organ pipe coral and the soft corals, sea whips, and sea fans) Pennatulacea (the sea pens)
	Scyphozoa		True jellyfishes and their relatives
	Cubozoa		The box jellies
	Hydrozoa		**Milliporina** (the fire corals) **Stylasterina** (the lace corals) + many other hydrozoans

Note: Names in boldface are groups that include species that produce calcified skeletons, and group names not in boldface are entirely soft-bodied organisms. All are marine organisms except for some hydrozoans. Only two of many fossil orders are shown, and most living orders of hydrozoan are also omitted.

wrenching them off the seabed using grappling hooks and hauling them to the vessel's deck, where they'd be dried and bleached in the sun, skeletal architecture was all that remained to sort one species from another, and biologists, being sometimes nerdy types, went to great lengths to describe and name the septa and all the other little bits and pieces.)

Today there are between 1,600 and 2,400 species of reef-building scleractinian corals, plus a similar number of non-reef-building deepwater species.[5] They share sixfold radial symmetry with four other orders of cnidarian: anemones, zoanthids, corallimorpharians, and black corals. These four groups all include some coral reef species, but none build calcium carbonate skeletons. Taxonomists group these five living orders with two important, but long extinct, orders of reef-builders, the rugose (order Rugosa) and tabulate corals (order Tabulata). (It is likely that the symbiosis between corals and single-celled algae, which has been so important to the success of reef-building scleractinian corals, first evolved in rugose or tabulate corals.) These seven orders are all grouped as subclass Hexacorallia because of their shared sixfold symmetry; they are more or less closely related to one another.

Four other types of stony corals occur on modern reefs: organ pipe coral, blue coral, fire corals, and the beautiful, delicate lace corals. Polyps of blue and organ pipe coral have eightfold radial symmetry, which puts them, along with the decidedly non-stony soft corals, sea whips, sea fans, and their relatives in subclass Octocorallia. Subclasses Hexacorallia and Octocorallia form class Anthozoa. The fire corals and lace corals, having very differently structured polyps, are placed in class Hydrozoa along with a number of small, non-reef-building forms, including a few that live in freshwater. Last, there are all the jellyfishes and the box jellies.[6]

So, corals are cnidarians that are mostly colonial, composed of numerous similar individuals called polyps. Each polyp is a simple cuplike creature with an upwardly directed mouth surrounded by tentacles. The polyp secretes an architecturally complicated skeleton of calcium carbonate upon which it sits cemented to the rock beneath, and taxonomists use details of this skeleton to identify the species. Adjacent polyps are connected to one another by living tissue so that the entire skeleton is inside/underneath the colony of polyps. As polyps grow, they periodically split vertically to create two sister polyps side by side, gradually enlarging the colony in this asexual way. Sexual reproduction leads to minute larvae that become new colonies. Most stony corals are scleractinians, but a few non-scleractinian corals also exist. And representative stony corals have lived in warm tropical seas for more than half a billion years.

Creatures that we would call corals first helped build reefs in the mid-Ordovician period (504 to 441 million years ago). Those rugose and tabulate corals played important but second-fiddle roles in reef building, on and off, from the Ordovician through the late Devonian, a period of about 150 million years, during which some of the largest reefs ever built were created.[7] For some reason, however, these corals then became less common, and they were pretty well gone by the end of the Permian.

The world's seas were not conducive to reef formation for 20 million years following the great end-Permian mass extinction. Once conditions improved by the mid-Triassic period (about 225 million years ago), reef-building began again. The rugose and tabulate corals were now extinct, superseded by the related Scleractinia, which had evolved from tabulates sometime during this interval. Although scleractinian corals were minor players at first, they rapidly became the dominant reef-builders and have remained so from mid-Jurassic times (175 million years ago) to the present. To put some of these time periods in perspective, human civilizations first developed agriculture about 8,000 years ago. Our species of human has been on the planet for about 200,000 years, and our genus, *Homo*, evolved from earlier hominids about 2.8 million years ago. Scleractinian corals have been building reefs for 175 million years, and reef-building by corals goes back almost 500 million years. We should treat the corals with some respect.

♦♦♦

The coral colony grows by polyp division (fig. 3). The division is lateral—that is, one polyp becomes two polyps side by side—and each species has particular rules governing how divisions of adjacent polyps are scheduled relative to each other. The result of these rules is that each coral species grows to produce a particular architecture that is visible in the orderly arrangement of polyps and at a larger scale in the overall shape of the colony. There is spectacular beauty in the fine-scale architecture created by these

Figure 3. The architecture of corals is varied and strangely beautiful. It is largely determined by simple rules dictating the sequence of divisions among adjacent polyps as the coral grows. Here, clockwise from lower left, are the regularly spaced polyps of *Diploastrea heliopora* off Zanzibar, Tanzania, a boulder-shaped *Colpophyllia natans* in front of a particularly laxly branched *Acropora cervicornis* off Bonaire, a series of cup- or plate-like colonies of *Echinopora lamellosa* cascading down a slope at Pemba Island, Tanzania, and in the lower right, a close-up view of *Acropora hyacinthus*, from the Keppel Group, Great Barrier Reef, photographed shortly after it had bleached. Images are used with permission as follows: *Diploastrea* and *Echinopora* both © A. J. Hooten, *Colpophyllia* and *A. cervicornis* © Robert S. Steneck, and *A. hyacinthus* © Ove Hoegh-Guldberg.

orderly patterns of growth—a phenomenon, incidentally, that is repeated time and again across the biosphere. The exquisite beauty of the nautilus shell, the fivefold symmetry of most starfishes and sea urchins, and the regularity with which leaves are arrayed on plant stems are each a result of the action of similarly simple rules about how to grow.

Patterns of colony growth are influenced, perhaps distorted, by the immediate environment. Some coral species exhibit quite different forms in deeper and in shallower sites. All coral species have their form modified when in a directional current, although they usually grow into rather than away from a current. The pace of growth is modified by light level, so a coral growing near an overhang will show faster growth on its outer, more brightly lit side. The presence of other coral colonies also affects growth, and corals will fight aggressively when they grow near enough to each other for their tentacles to reach across and fire their toxic stinging nematocysts at one another.

In the 1970s, marine biologist Judy Lang revealed in a series of field experiments that there was a hierarchy in competitive abilities among Caribbean coral species.[8] Certain species would reliably outcompete certain other species, meaning that, when they grew close enough to one another, the competitive dominant would succeed in killing the adjacent polyps of the less dominant species, thereby preventing any further growth. In that way the competitive dominant could expand a bit more, while the less dominant species would cease growing on that part of its diameter. Lang also showed that, to a remarkable degree, fast-growing species of coral tended to be competitively inferior to slower-growing species, although they could sometimes overtop the slower-growing species and shade them out. Of course, all this fighting, this thrashing about at one another with poison-tipped barbs propelled explosively from nematocysts, happens at a leisurely pace, and the supplanting of one coral by another takes months or years to complete.

So far, I have dealt only with coral growth. The processes involved lead to a species-typical colony architecture. The interactions of adjacent colonies with each other, and with other aspects of their environment, help determine reef architecture at a larger scale. Yet even though we call them coral reefs, corals are not the only creatures producing calcium carbonate. The great majority of reef invertebrates build their skeletons out of this material, and although many of these are not attached to the rocks, their skeletons add to the carbonate mass once they die. One way of looking at a coral reef is to see it as a rich community of organisms, most busily engaged in manufacturing solid calcium carbonate using calcium ions and carbonate ions dissolved in the water around them. Sponge and starfish spicules, sea urchin tests (hard shells), mollusk shells, crustacean carapaces and claws, the hard bits in worms of many different kinds, and even the tests of tiny, single-celled foraminifera all contribute calcium carbonate along with traces of other compounds to the mass of solid material that is being built through the lives of these animals. The coralline algae play a major role in reef building because they tend to grow as veneers across rocky surfaces, cementing everything together. Reefs are *coral* reefs because the corals are the primary contributor of carbonate rock to their structure, but whereas reef rock is all biogenic (formed by organisms), it is definitely not all of coral origin. In some parts of the world the outermost, shallowest part of a reef is a solid pavement so dominated by coralline algae that geologists refer to the "algal ridge" in naming it.

♦♦♦

While all this construction is going on, equally potent forces of destruction are at work. First come the physical forces, particularly the big storms, which topple and overturn coral colonies and break up extensive thickets of branching coral to create rubble beds. These rubble beds can then persist for years because the individual bits of rubble keep rolling around, making it difficult for new corals to settle and grow or for coralline algae to cement the rubble in place. Sometimes a storm will cut a passage through the high reef crest that typically exists at the outer margin of a reef. When that happens, the drainage pattern during falling tides is altered, affecting the shallow reef flat and the lagoon behind the reef. Coral that was previously always submerged, even on the shallow reef flat, is suddenly exposed to the air and killed.

In the early 1970s, I made a field trip to Heron Island on the southern Great Barrier Reef a couple of weeks after a cyclone (a southern hurricane). The devastation was stunning. On some shallow parts of the reef, I saw large coral colonies that had been broken off and flipped. In other places new channels had been scoured and adjacent colonies had newly abraded, polyp-free surfaces, the result of sand carried in the water. Nor is storm damage limited to the shallowest places. When Rob Day, a graduate student from Sydney University, dived to his twenty-meter-deep study site to check on tiles he had attached to a large coral boulder, he found that the boulder was no longer there.[9] Or, rather, the two-meter-high boulder had been completely buried and killed by new sediments carried down from the reef above. That boulder was at least five hundred years old when it was killed, and Rob's experimental tiles were buried with it.

Prevailing currents and tidal surges also set patterns of water flow around and through reefs, influencing how corals grow and determining the form of the reef. Along with storms, currents and tides shape the large-scale topography of the reef. Another major physical force is sea level change, either because the overall sea level is actually rising or falling or because locally there is emergence or subsistence of the rock. Changes in sea level are almost always very slow; reefs mostly keep up with rising sea levels, growing upward to stay close to the sea surface, but they are progressively killed by emersion during low tide in times of falling sea levels. When sea level changes rapidly, as when an earthquake results in vertical faulting or in times of rapid climatic change, entire reefs can be killed by becoming elevated high above the high tide level or being too deeply submerged: the algal symbionts of reef-building corals need sufficient light for photosynthesis, and coral growth is greatly slowed when photosynthesis is not occurring. Some regions of the tropical Pacific Ocean include numerous sea mounts—flat-topped mountains rising to within several hundred meters of present-day sea levels, with evidence of fossil coral reefs on the upper surface. These coral reefs flourished during times of peak glaciation within the Pleistocene, when sea level was much lower than now, but were killed when sea level rose rapidly during one of the interglacial episodes.

Also important in curtailing the growth of reefs is a veritable army of organisms ranging from large parrotfishes to tiny worms and sponges that are collectively termed bioeroders. These creatures scrape, bite, carve, excavate, burrow into, and/or dissolve carbonate rock, feeding on the algae and bacteria that live on rocky surfaces or within the surface layers of the rock itself.

Bioerosion plays a major role in shaping the fine-scale structure of a coral reef.

Boulderlike massive corals of a number of species, such as the one Rob Day lost in that cyclone at Heron Island, grow as a more or less spherical or domelike mound, slowly increasing in diameter year by year. The outer surface of such colonies is covered by living polyps, while underneath them are the accumulated skeletons of all their ancestor polyps. However, most massive corals are not the solid structures that you would expect once they reached sizes more than 30 centimeters or so in diameter. Instead, they are hollowed out to greater or lesser degrees by the burrowing activities of worms and mollusks and the dissolving activities of acid-secreting sponges to create internal cavities of complex shape and size that, if they connect to the outside, as they often do near the base, become important cave habitats occupied by a myriad of other creatures, especially fish, crustaceans, and echinoderms.

Scraping at outer rock surfaces by herbivorous bioeroders such as sea urchins and parrotfishes can be similarly extensive. I think of parrotfishes as living sand-manufacturing machines. You can hear them biting and scraping as they move about over a reef feeding on algae living on and in the surface layer, and they leave plenty of scrape marks on the rock in their wake. They also engage in a near continuous process of defecation, with visible streams of extruded material floating down from their vents. This material is mostly sand resulting from all that rock scraping, plus some undigested algae. In 1982, Ross Robertson, of the Smithsonian Tropical Research Institute, published his observations of feeding in parrotfishes on a Panamanian reef. He recorded the consumption of defecated material by other fishes and calculated that the average fecal pellet expelled by a parrotfish would be consumed seven times by other fishes before it finally reached the substratum a meter or so below the parrotfish that started that particular chain of waste recycling. In 2003, Dave Bellwood and colleagues at James Cook University in Queensland used observations of feeding by the world's largest parrotfish, *Bolbometopon muricatum*, to calculate the amount of rock material removed by its grazing on the Great Barrier Reef. They estimated that a single fish 1.2 meters long and weighing 45 kilograms can annually excavate

5.7 metric tons of carbonate rock and convert it to sand. Given the abundance of this species on the Great Barrier Reef, they estimated the rate of rock converted to sand by *Bolbometopon* grazing to be 279.3 metric tons per hectare of reef per year. (This species is by far the most important bioeroder among parrotfishes.)[10]

So, a coral reef is a process—or perhaps a set of competing processes. It exists as a fragile balance between the forces of reef building and the forces of reef erosion. Whether the reef is growing or degrading depends on which forces are dominant at any given time. Most coral reefs examined over the past fifty years or so have been found to be at near equilibrium between growth and erosion, and usually slightly on the positive (growth) side of that balance point. In recent years, however, a growing number of studies report negative reef growth, with the forces of erosion being more powerful than the forces driving reef building.

♦♦♦

Up till now, I've been talking about the day-by-day construction or destruction of reef rock. In the grand scale of space and time, coral reefs are imposing structures, majestic masses of carbonate rock including fine-grained limestones and coarser, in situ reef deposits. They are the largest structures built by living organisms on this planet. Long before HMS *Beagle* rounded Cape Horn, long before he had seen a coral reef, Charles Darwin was pondering their formation. He knew that reefs could be divided into three types: fringing reefs formed along rocky shores, barrier reefs bordering deep oceanic waters but separated from land by a protected lagoon, and atolls existing at midocean locations far from any continent. He also knew that corals did not grow in deep water, so these three types of reef were, in turn, relatively easily understood, somewhat more problematic, and very hard to explain. How indeed could creatures that did not live in water much more than 30 or 40 meters deep build a more or less circular reef, with occasional low, sandy islands dotted with palm trees, all surrounding a tranquil lagoon, in oceanic waters otherwise many kilometers deep? His finding of fossil corals in the Chilean highlands reminded Darwin that landmasses could shift vertically and led to his hypothesis explaining the existence of atolls, barrier reefs, and fringing reefs as all caused by coral growth around the edges of slowly subsiding landmasses. This was the idea I somehow learned about as a child in Bermuda.

Once he reached reef waters in the Pacific, Darwin made observations confirming the shallow depth limits of corals. Subsequently (in 1842) he set down his hypothesis and explanations in *The Structure and Distribution of Coral Reefs,* his second most important book.[11] This hypothesis stimulated a century-long quest by geologists to solve what became known as "the coral reef problem" by drilling down through reefs to discover whether they would confirm Darwin's expectations. Do coral reefs have a deeply buried basalt platform that was originally at sea level but had subsided over time?

As often happens in science, the various efforts to test Darwin's hypothesis on reefs around the world revealed that he was correct, more or less, much of the time, but also that the development of coral reefs is a much more complex process than Darwin suspected. The processes of reef growth and reef erosion take place in an environment that has not always been the environment of today. These environmental changes have been profound over the span of geological time, and no coral reef on the planet has grown continuously since the first appearance of rugose corals, or even since the first scleractinian corals in the mid-Jurassic. Reef growth has been interrupted often, sometimes for millions of years, and there are multi-million-year gaps in the global record as well—times when ocean chemistry or climate was not favorable for reef building anywhere on the planet. During these hiatuses, a few corals persisted as rare creatures, incapable of growing quickly enough to build substantial reefs. When conditions improved, reef construction resumed or started anew in other locations. When reefs are present, subsidence of an underlying platform can be important, as Darwin hypothesized, but climatic shifts, shifts in ocean chemistry, alterations in sea level, and movements of crustal plates through plate tectonics also influence whether a site is suitable for reef growth. In the relatively recent past, the Pleistocene was a time marked by at least twenty substantial fluctuations in sea level from around 120 meters below present day to 10 meters above. These alternately submerged and elevated existing reef platforms. We can see evidence of this much more complex development of reef structure in the cores drilled to test Darwin's ideas.

For example, in 1952 a bore hole drilled on Enewetak Atoll, one of the Marshall Islands in the southwest Pacific, yielded 1,400 meters of carbonates before reaching a basalt base, a result that could only occur if the reef was growing on a subsiding platform. The earliest carbonate rock dated to mid-to-late Eocene times 45 million years ago. Later work analyzing this, and several subsequent cores, revealed numerous discontinuities in the limestone best understood as periods when the reef was elevated above sea level and being eroded away.[12] One such discontinuity lasted 6 million years during the mid-Miocene. The development of Enewetak has not been the slow, continuous process Darwin imagined.

The Hawaiian Islands form a linear chain that extends 2,500 kilometers northwest from Hawai'i to Kure Atoll and Midway Island, a result of sustained volcanism from a mantle hotspot under a slowly northwesterly moving Pacific plate (fig. 4). Hawai'i is the youngest island, still volcanically active and with the highest elevations, while the youngest basalt at low-lying Midway Island was extruded 27 million years ago. The chain continues to the northwest with a series of still older basaltic seamounts, and southeast of Hawai'i, the Lō'ihi seamount already rises 3,000 meters from the ocean floor, the newest volcano in this chain. On Hawai'i, Mauna Kea, at 4,200 meters elevation, is currently the highest point in the chain, although erosion and subsidence reduce that elevation by about 4 millimeters per year. A chain of progressively older volcanic islands, slowly subsiding—surely the Hawaiian Islands should be another poster child for Darwin's hypothesis!

Again, that is not quite the case. Across these much younger islands, there is nowhere the considerable depth of carbonates evident at Enewetak.[13] Although coral reefs occur around all the Hawaiian Islands, volcanic basalts are evident everywhere, and carbonates never form other than a thin veneer on top of basalt. The development of Hawaiian reefs has been driven primarily by fluctuations in sea level during the Pleistocene that alternately elevated and submerged shores, and by seasonal periods of strong seas that have been so effective at stripping away coral growth that there has been essentially no accretion to reef platforms for the past 5,000 years. On O'ahu, for example, the fossil Waimanalo reef dates to the last Pleistocene interglacial period 134,000 to 113,000 years ago while the Wai'anae reef is 247,000

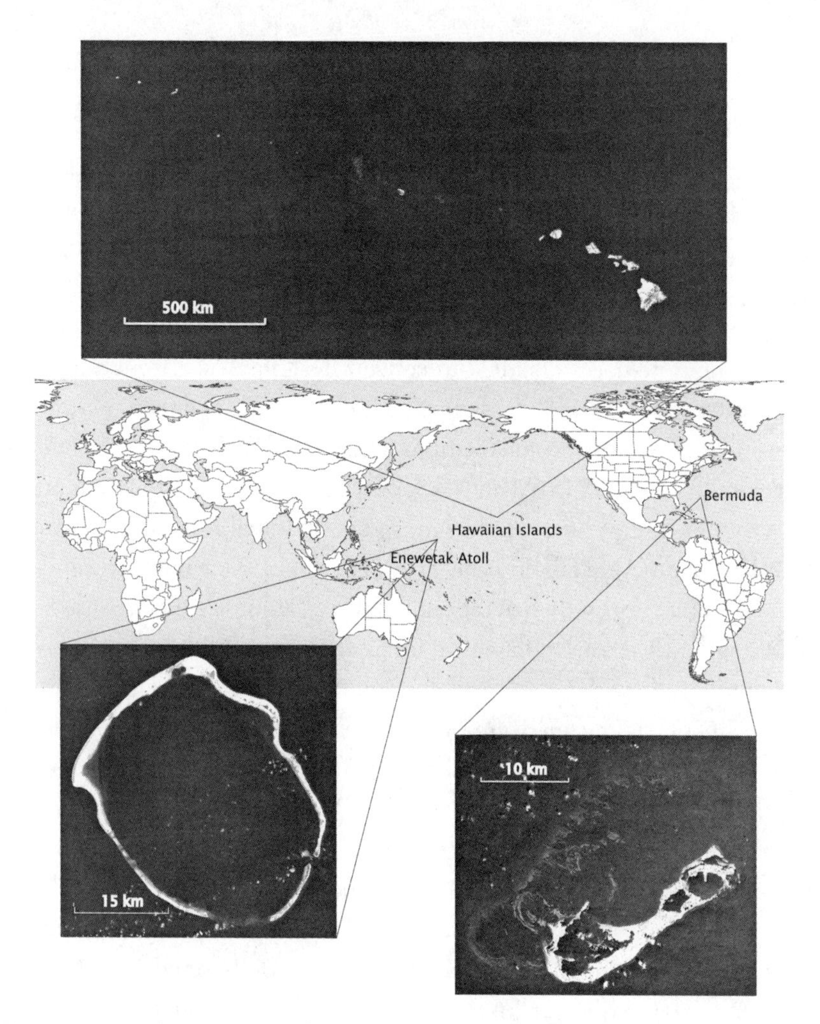

Figure 4. Bermuda, Enewetak Atoll, and the Hawaiian archipelago are shown in three satellite images. Enewetak, approximately 40 kilometers across, is somewhat larger than Bermuda, 24 kilometers in maximum length. Both are limestone islands supported on a basalt base in midocean depths, but Enewetak is substantially older, with a much thicker limestone cap. The Hawaiian archipelago is a linear series of basaltic peaks arising from midocean depths that exist as large islands with volcanic peaks in the younger southeastern region but are subtidal with limestone caps in the northwest. The chain is approximately 2,100 kilometers long from Kure Atoll to Hawai'i. Figure assembled using four files from Wikimedia Commons. These are (1) base world map developed by John Tann and supplied under Creative Commons Attribution 4.0 International license; (2) Image of Bermuda courtesy of ASTER/JPL/NASA; (3) Image of Enewetak Atoll courtesy of NASA/USGS, sourced originally from a NASA/USGS Landsat 8 satellite image of Enewetak Atoll captured on February 10, 2014, archived at http://glovis.usgs.gov; and (4) Image of the Hawaiian Island chain supplied as a composite satellite image courtesy of NASA.

to 200,000 years old. There is no evidence of reefs earlier than this. Cores of the O'ahu platform seldom yield accumulations of carbonate rock more than 30–40 meters thick, and much of this material is in the form of sands and reconsolidated limestones with much evidence of erosion during periods when the material was above sea level. The environment has not been conducive to the construction of fossilized, in situ reefs many meters thick; rather, the system has operated to convert dead coral skeletons and other carbonates to sand, which is often recemented as fine-grained limestones.

Bermuda lies well outside the tropics (fig. 4). Its reefs are the most northerly coral reefs in the world and exist only because the Gulf Stream brings warm water up from the Caribbean. In age of initial formation, Bermuda falls between Enewetak and the Hawaiian chain. Volcanic activity commenced in the Eocene, when the main bulk of the subsea platform was built. The final push in early Oligocene times about 33 million years ago saw a mountain rising perhaps 1 kilometer above sea level. This mountain did not subside to any significant degree, as I had understood as a child, but instead was eroded away by wind and rain during subsequent years, much as is now happening in the Hawaiian chain. Volcanic basalts, remnants of that eroded mountain, are now at least 60 meters below present-day sea level and buried by a thick but young cap of reefal limestones deposited once sea level rose and climate warmed (late Pliocene or Pleistocene, 2 to 3 million years ago).[14] The islands of Bermuda lie primarily along the southern edges of two calderas in the basal volcano but did not develop simply through uplift of reefs as I had understood. Most of the rock above sea level is consolidated sand dunes (eolian limestones) formed through erosion of reef as its rock is worn away by wave action, by storms, and by the active grinding, drilling, and chewing of bioeroders. The alternation of glacial and interglacial episodes in the Pleistocene ensured that reef development was frequently interrupted. During cold periods, when North American glaciers were at their maximum, the low sea levels allowed for lots of reef erosion and winds built immense dunes—the beaches were probably even more magnificent than they are today. Ultimately, those dunes became consolidated into the fine-grained limestone that forms the topography and has been quarried and sawn into the flat tiles used to build the distinctive roofs of Bermudian houses.

In contrast to Enewetak, the Hawaiian Islands, and Bermuda, the Great Barrier Reef, or GBR, is an immense set of reefs resting on a continental shelf (fig. 5).[15] Surely it is an immensely old structure that resulted from gradual subsidence of this shelf as Darwin proposed? Thanks to the efforts of Peter Davies and colleagues at Australia's Bureau of Mineral Resources, we know quite a bit about the history of One Tree Reef, part of the Capricorn Group at the southern end of the GBR, and of the GBR more generally. Rather than a story of slow subsidence, it's one of plate tectonics, sea level rise and fall, runoff and turbidity, and coral growth and destruction interacting serendipitously through long periods of time. This history highlights many of the features important in the development of coral reefs around the world.

One Tree Reef did not exist in the deep past because Australia's Queensland coast was not in the tropics back then and water temperatures were too low. Lying 70 kilometers offshore and about 20 kilometers from the edge of the continental shelf, the One Tree Reef of today is a small (3-by-5-kilometer) platform resting on a more or less flat plain about 60 meters deep—it is perhaps the fourth iteration of a reef that has formed and reformed at this location in a very short 400,000-year span since the mid-Pleistocene. Present-day One Tree Reef began forming about 8,000 years ago, building on the high points of a limestone mesa above a coastal plain that had been high and dry and extensively weathered by wind and rain for some 80,000 to 100,000 years. But let's begin at the beginning.

What is now the northern Great Barrier Reef first entered the tropics only 24 million years ago, in the early Miocene, as Australia drifted north at a leisurely 7 centimeters per year. The southern Queensland shelf, and the future One Tree Reef, did not reach the tropics and the climatic conditions that might favor reef growth until mid-Pliocene times about 2.5 million years ago.

While Australia was drifting north, its continental shelf and the future Great Barrier Reef were separated by the deep Townsville Trench from two large, shallow plateaus, Queensland and Marion, which together formed the southern edge of the Coral Sea Basin. New Guinea and islands to the east marked the northern boundary of this basin. Sea level also was changing. In

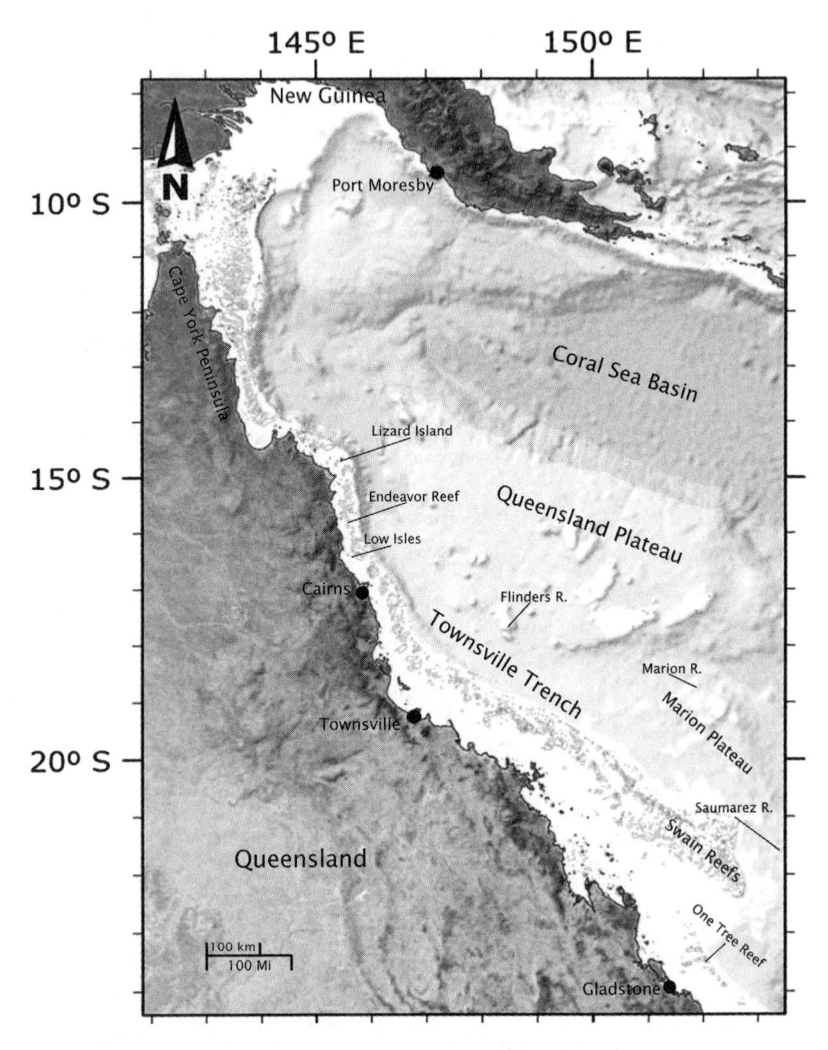

Figure 5. The Great Barrier Reef in relation to other marine structures and to New Guinea and the Australian continent. Reefs mentioned in the text are named. (One Tree Reef and surrounding reefs are better seen in fig. 7.) Base map courtesy of eAtlas and Australian Institute of Marine Science.

the early Miocene, 24 million years ago, sea level was low and both plateaus were shallow enough to support extensive coral reefs. However, Australia's continental shelf was dry land being inundated by riverine sediments from the highlands to the southwest. Global sea level continued to fall slowly, until it reached a low point 10.5 million years ago. Local subsidence of the two plateaus was insufficient to prevent the Marion plateau also becoming dry land as sea level fell but initially kept the Queensland plateau suitable for reef growth. Continuing subsidence over the next several million years, to about 3.5 million years ago in the mid-Pliocene, finally took the Queensland plateau below the photic zone, and reef growth ceased except for a few isolated reefs on its southern margin, such as Flinders Reef, that remain even today. The subsidence put the Marion plateau back underwater by 5 million years ago, but initially cooler waters and muddy sediments derived from the eroding uplands made it unsuitable for coral growth until the early Pleistocene. Reef growth, notably at Marion Reef and Saumarez Reef, commenced then and has continued, with several interruptions, until today.[16]

During all of Miocene time when reefs flourished on one or both of the plateaus, the continental shelf of Australia was without reefs and mostly above sea level. During the first half of the Miocene, from 24 to 15 million years ago, the northern shelf was within the tropics and sea level rise had initially submerged it, but it was a site of extensive riverine and deltaic sedimentary deposits, and the seas at this time were high in nutrients. Neither factor made it a good place for reef growth. Then with global sea level falling, the shelf again became emergent and remained so until 5 million years ago; there was extensive erosion and sediment transport toward the shelf edge. During Pliocene times (5 to 1.8 million years ago) and into the Pleistocene (1.8 million to 11,000 years ago), sea level rise and local subsidence put the shelf back into appropriate depths for reef growth, but the southern portion of the shelf did not even reach the tropics until about 2 million years ago (remember, Australia was moving north all this time).

The newly submerged continental shelf continued to be subject to intense riverine and alluvial sedimentation throughout the first half of the Pleistocene, and it was not until about 452,000 to 365,000 years ago that growth of the Great Barrier Reef began in earnest. Sea level had been fluctuating continuously by around 120 meters as continental glaciers waxed and waned throughout the Pleistocene, and during the final 400,000 years there were four such cycles. At each low stage, the shelf was emergent, and wind and rain reshaped elevated reefs. Then, with sea level rising, the shelf again submerged and reef growth began again on the eroded surfaces of what had been built before.

The final low stand of sea level during the Pleistocene occurred about 19,000 years ago, when sea level was 125 meters lower than today. Queensland's continental shelf was again high and dry. With the warming of climate and glacial melting, sea level first rose rapidly and then more slowly, reaching present-day sea level or slightly higher at 6,500 years ago. The rate of rise was as much as 50 millimeters per year and averaged 10 millimeters per year during these 12,500 years, about three times faster than at present. Reef growth, as usual, was delayed following resubmergence, no doubt due to the need for clear, nutrient-poor water. Despite its enormous size, the Great Barrier Reef is little older than the much more limited reefs of Hawai'i and much younger than the reefs of Bermuda or Enewetak.

At One Tree Reef, and elsewhere throughout the region, reef growth recommenced about 8,000 years ago. The reefs grew vertically, lagging behind the rising sea level but catching up to sea level around 5,000 years ago. Vertical growth rates were about 5 meters per 1,000 years, and both reef and sea surface reached current sea level about 4,500 years ago. With vertical growth halted, reefs grew laterally by growing slowly into currents on the windward side while expanding more rapidly to leeward through the erosion and deposition of carbonate sediments in lagoons and on backreef slopes. This period of lateral growth lasting a scant 4,000 years or so has built the One Tree Reef of today, with its diversity of habitats from steep windward slopes through algal ridge and extensive reef flat, a sand-floored lagoon, and a backreef flat and slope to the surrounding plain. The overall shape of One Tree Reef has been set by the shape of the Pleistocene mesa on which it grew, a mesa composed of the remains of about four successive Pleistocene reefs. To put things in context, while the crest of One Tree Reef stands 40 meters above the immediate shelf platform and 60 meters above the typical shelf depth in this region, the modern One Tree Reef is a layer of reef-derived limestone just 10 to 12 meters thick, deposited over 8,000 years, on top of a Pleistocene precursor some 25 to 35 meters thick. If we date the modern reef from 4,000 years ago, when it assumed its modern form, it is no older than the pyramids of Egypt! Admittedly, humans built the pyramids over decades, while One

Tree Reef was already 4,000 years old when it reached the surface and has continued developing until the present.

Living reefs at Enewetak, Hawai'i, Bermuda, and the Great Barrier Reef are very different, and yet they are the same. Each reveals a process of growth far more complex, far more erratic, than Darwin's hypothesized slow upward massing of carbonate rock as generations of coral polyps tirelessly built their skeletons on a gradually subsiding platform over immense spans of time. Instead, coral reefs are dynamic creations that change a lot through time under the influence of multiple agents beyond the polyps. Some of these agents are other forms of life, others are physical and chemical processes of the marine environment, and still others are grand movements of tectonic plates and massive shifts in climate. Despite the many agents working, often at cross purposes, coral reefs have a beauty and a correctness of form that I find amazing. They are appropriate to their location much like the very best of human architecture, and this is achieved without any real architect in charge. Reefs are majestic structures that are built, modified, partially demolished, and rebuilt, on and on, yet they always look appropriate. They are never completed until, for one reason or another, the corals that are building them die out at that location. Charles Darwin never suspected this dynamic existence; it means that reefs have dramatic histories.

Those histories are also far shorter than we might have expected. True, most living reefs have been around for far longer than any of our cities (the only other biogenic structures that rival them in size), but none extend far back into the Pleistocene without interruption, and some of those interruptions have been millions of years long. The Great Barrier Reef, the largest reef province on the face of the planet, one of the largest in Earth's entire history, has been built several times, disassembled, and rebuilt. What we see today was built over a scant 8,000 years of history. The building of coral reefs is far more amazing than Darwin ever dreamed.

THREE

Ever Wonderful, Always Different

We live in an increasingly strange world. Although most people on our planet have never seen a coral reef, we have all seen images, even videos, of coral reefs. Paradoxically, reefs are so well known that they can become the environment for animated movies like *Finding Nemo*. We all know a reef when we see it.

And yet, as I pointed out in chapter 1, we don't. A photo of a coral reef is not a reef, no matter its level of spectacle or the skill of the photographer. It is this situation of unknowing knowing that makes it very difficult to convey just how utterly amazing coral reefs are. Our access to visual representations of them has deadened our awareness of what a coral reef is. We have seen lots of images, but we do not have the context, the essential knowledge about coral reefs, to comprehend what it is we are looking at.

The last chapter revealed the coral reef as a way of becoming: a continuing process of growth and destruction spun out over long periods of time, driven by multiple agents, some physical, some biological. We call the result a reef, not recognizing that it is still unfinished. These massive calcareous structures are the largest structures built by living organisms on this planet. In their origin, growth, senescence, death, and rebirth, coral reefs tell a tale of dynamic change, full of sudden reversals, rather than a story of slow, steady growth over eons of time. That dynamism is a first hint of their fragility, their impermanence, their sheer improbability—these are not simple entities, but neither is their improbability the only thing amazing about them.

How can I convey the wondrousness of living coral reefs? More pretty pictures just won't do it. I find something about the experience of being on a reef that is far deeper. These are special places, and a visit yields long-lasting memories of wonder, elation, and exuberance. I begin with a story from the southern Great Barrier Reef.

♦♦♦

I remember well my first trip to One Tree Reef. I had been in Australia about five years by 1973 and had been doing field research out of the Heron Island Research Station just over the horizon to the west of One Tree Island.[1] In my own mind I was becoming an old hand at coral reef science, although I had still seen relatively little of the Great Barrier Reef, never mind reefs farther afield. In those days, few of us had a really wide geographic experience of reefs, but I was spending about three months a year at Heron Island.

Heron Island is a typical Great Barrier Reef sand cay—built entirely of reef-derived carbonate sands, surrounded in part by beach rock (lithified beach sands), located on the leeward end of a reef platform, and ephemeral on a geological timescale but sufficiently permanent in many cases to include a freshwater lens and vegetation including mature trees. Heron Island possesses a dense forest of Pisonia trees (*Pisonia grandis* is a large-leafed, soft-wooded tree that reaches 20 meters in height and dominates sand cays of the Western Pacific). These trees are home to many thousands of black noddies (*Anous minutus*), which festoon their branches while being serenaded seasonally by the mutton birds (wedge-tailed shearwaters, *Ardenna pacifica*) caterwauling at the mouths of their burrows beneath (fig. 6). There are some herons as well.

Despite its name, the Heron Island Research Station was not what those words might conjure in the mind of a scientist who'd grown up in North America. Established in 1954, it consisted of several basic living huts, a shared kitchen and dining building, a laboratory building with one large room with bench space for twenty to thirty students and a couple of much smaller rooms, a system for pumping sea water (at high tide only because the intake was too shallow) to an aquarium facility that had just one functional aquarium, and essentially none of the equipment a marine laboratory

Figure 6. One of the many thousands of black noddies that live within the Pisonia forest at Heron Island. This one is perched proudly behind his particularly slipshod nest. Noddies are one of the two avian reasons why I much preferred One Tree Island to Heron Island. Photo © Peter Sale.

might be expected to possess. Its sole boat, the dory, was a small, high-sided, heavy wooden boat of dubious age, powered by a single-stroke, inboard gasoline engine that was started by rotating its flywheel as rapidly as you could manage using your bare hands. It putt-putted rather than screamed across the seas and was as unsuitable a small boat for diving as I could imagine. (If you did not remember to hang a securely attached line over the gunwale before entering the water, there was absolutely no way of getting back into the boat, because you had to tie your tank to this line, throw mask and fins high into the air over the gunwale, and clamber in using the rudder as a step.)

On the other hand, the research station was next to a seemingly endless (in the absence of a useful vessel) reef waiting to be explored. There is something rather strange about coming all the way to a coral reef only to then grumble about the lack of a well-equipped lab, and I did not grumble for long. The station's deficiencies were a blessing in disguise, forcing me to get outside and learn.

Indeed, I had been quickly able to solve its one serious deficiency by using university funds to secure my own, more appropriate, dive boat. After I convinced my skeptical superiors at the University of Sydney that I was capable of operating a tiny research vessel in faraway Queensland, I moved it to Heron Island, where it spent the months when I was not in the field upside down under a Pisonia tree—it was a tiny 3.5-meter-long boat.

One of the unexpected joys of spending time at Heron Island was that I got to meet other scientists passing through or doing more extensive research there. Sharing a kitchen and dining room tends to break barriers that might otherwise persist, and I had early on developed the habit, when booking time at Heron Island, of inquiring who else was likely to be there at the time. For all its deficiencies, the Heron Island Research Station was the only real research facility on the Great Barrier Reef. Either you did field research by expedition or you went to Heron Island. For scientists visiting Australia, a visit to Heron Island was an inexpensive way to see the reef; to go there, you had to have a research project you were working on, and some projects were tissue thin, but they gave scientists a chance to experience the reef.

Heron Island was not quite the only reef research facility, however. Among the people regularly passing through was a motley collection of researchers, chiefly but not entirely from the Australian Museum (also down in Sydney), who were conducting research at nearby One Tree Island. If the Heron station was basic, One Tree Island was less than that. During my first couple of years in Australia, there was nothing on One Tree Island at all. Research teams from the museum or from the University of New England (Americans, note that this is an Australian university, in Armidale, New South Wales) would pass through Heron, spending one night at the station en route to or returning from One Tree.

The Armidale group comprised several graduate students and technicians led by Harold Heatwole, a bizarrely amazing academic of American origin whose specialties were the ecology of sea snakes and of remote island communities. In the near absence of snakes, his island interests brought him to One Tree. He was interested in the processes of colonization and extinction that take place continuously on small islands. The types of species that colonize, the length of time before they go locally extinct again, and the overall composition of the biota provide an endless set of interesting questions for the academic mind, and Hal Heatwole was totally tangled up in this web of fascination.[2] For him, One Tree Reef was just the scaffolding that supported the island, a tiny 4-hectare speck of dry land built not of sand but of coral rubble, surrounded by ocean, with the nearest other land, Heron Island, 15 kilometers to the west and the Queensland coast 90 kilometers to the southwest.

Located toward the windward face of its reef, this rugged island, with a small, brackish pond in its center, supports a large clump of screw palm, *Pandanus* sp. (the "one tree" of the island's name), near the pond, as well as three or four small clumps of stunted Pisonia trees around its edge. The remaining vegetation is mostly a sparse mixture of herbs, mainly the succulent *Sesuvium portulacastrum* and the shrub *Argusia argentea.* The island fauna includes insects and other terrestrial invertebrates, including some memorably large centipedes with a predilection for crawling into wet dive boots, and nesting seabirds of several species. A large, 2.5-meter-tall eagle nest stood within the *Pandanus* clump on January 10, 1843, when the island was first visited by a white man. J. Beete Jukes, naturalist on board HMS *Fly*, was surveying the islands and reefs of the Great Barrier Reef and Torres Strait. He reported the "one" tree and the nest. That nest, actually an osprey nest, is still present but rarely used.

One Tree Island fronts onto the main lagoon of One Tree Reef, a lagoon that becomes ponded almost a meter above sea level during low tide. Because of this, travel to One Tree Island has to be timed to the tide, hence the overnight stays at Heron Island.

The reef itself has been the main attraction for most scientists who have worked at One Tree, including Frank Talbot, then director of the Australian Museum and a reef fish ecologist by training. In 1965, he had commenced annual expeditions to One Tree Island to use it as a base for detailed taxonomic and ecological study of the coral reef by museum staff and collaborators from Australia and overseas. Initially, they used a standard expedition mode of operation, bringing large quantities of gear including tents, food, water, and everything else they would need for a stint of several weeks' field research. This was the same approach the Armidale group used, although the museum groups tended to be larger, and rowdier when not working, than did Hal Heatwole's group of island ecologists. In 1969, the Australian Museum secured a long-term lease on the island from the Queensland government and erected three huts—the beginnings of a permanent facility. Little more than hard-walled tents, built of corrugated iron roofing on timber frames, with window openings protected by plywood shutters, and with double-deck bunks part of the structure, these huts had no floors, just smoothly raked rubble. They were fastened down by baling wire guywires from the eaves to large stakes pounded into the surrounding rubble, and none were ever lost to cyclones. Those guywires led to some interesting, potentially dangerous, stumbles by scientists wandering about at night.

With the establishment of quasi-permanent structures, it became necessary to have somebody live permanently at One Tree Island, and a succession of couples have had that experience from 1969 to the present day. I got to know these caretakers, because they would visit Heron Island weekly for supplies or to pick up or drop off researchers. And so, after meals, coffee, and beers together over a couple of years, the time came when I was invited to visit.

On one of my trips to Heron Island in 1973, Ted Chilvers, a fiftyish, jack-of-all-trades Aussie and One Tree Island caretaker, drew me a rough map with the landmarks I'd need to cross the reef when I came over in a couple of days (fig. 7). It was to be an "over one day, back the next" chance to see the place. I did not know at the time that plans were afoot to transfer the lease to the University of Sydney—I was being tested to see if I'd like to work there, because as the university's most active field researcher working on the reef, my working there was all part of the plan being quietly hatched around me. (Younger scientists may marvel at the top-down paternalism of Australian universities back then, but that was the social environment in which we all lived.)

The day for my visit, chosen because it was close to the peak spring tide and thus one in which crossing the reef would be easiest, dawned warm and

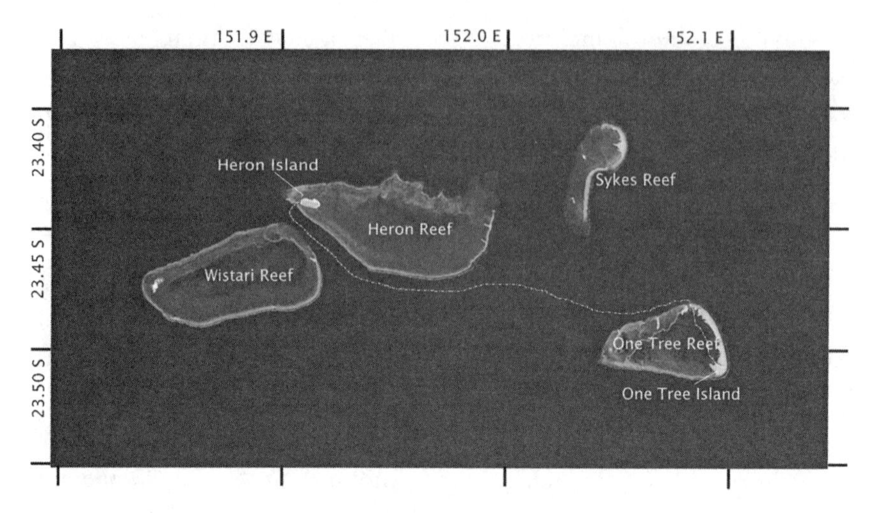

Figure 7. Satellite view of Heron and One Tree Reefs, and the path I took to visit One Tree Island. Heron and One Tree, along with several other reefs and islands, comprise the Capricorn Group, which with the nearby Bunker Group forms the southernmost portion of the Great Barrier Reef. Landsat 8 image, June 19, 2018, courtesy of USGS via http://glovis.usgs.gov/.

blissfully calm. While the ocean is almost always moving, sometimes violently, there occasionally occur these eerily calm, still days when the water is like glass reflecting the sky. This was one of those days.

As my technician, Rand Dybdahl, and I set out from Heron Island in our tiny boat, quickly crossing the 500-meter-wide reef to the deeper water along its southern face, we paused long enough to watch for a few minutes as small butterflyfishes, damselfishes, and wrasses swam about upside down just below the water surface and 10–15 meters above the reef where they belonged. While we saw reflections of the clouds, they were seeing a reflection of the reef below.

Over the years, I've seen fish do this a number of times. They take an enormous risk when they swim up far above the substratum, because there are plenty of larger fish about, and small fish away from cover are inviting trouble. Swimming to the mirror world on the surface—a bit, I suppose, like passing through the looking glass—reveals just how strongly fish rely on vision to orient themselves in space. Their inner ears tell them the difference between up and down. Their swim bladders tell them if they are ascending or descending to deeper water. They know they have traveled far from the place they usually occupy to this new place. And yet, they let the false evidence from their eyes bring them up to the surface, often swimming upside down for minutes at a time.

Fish will sometimes swim upside down in caves, if there is a sandy floor and sunlight from the cave mouth. In this case, it is the direction from which the light appears to come—up from the floor—that causes the inverted posture. One of the main reasons for a feeling of nausea when suffering seasickness is because of the contradictory messages coming from one's eyes and inner ears. That's why going on deck and watching the horizon can be beneficial—watching the horizon you realize that your body and the vessel itself are being tossed and turned, and suddenly your eyes and inner ears are no longer in disagreement. They are shouting in unison that your universe is going up, down, sideways, and around the corner; it is still uncomfortable, but less nauseously so. Makes me wonder whether fish feel any sense of discomfort while swimming about, high above the reef, upside down, on a calm day.

In any event, we left the confused fish behind once we moved out to slightly deeper water and followed the line of Heron Reef to the east. After about forty-five minutes we reached the eastern end of the reef. Heron Island had just about disappeared beneath the horizon, and we were alone, in a small boat, on an immense circular sea. That feeling did not last long, however, because One Tree Island soon appeared as a smudge on the horizon ahead of us. Moving through the slightly turbulent waters east of Heron Reef, south of Sykes Reef, and west of One Tree Reef, we proceeded along the northern side of One Tree Reef until we reached the designated part of the reef (as judged by locations of a couple of rubble banks on the crest—this was decades before GPS) for our crossing to the lagoon. Moving in closer to the reef edge, Rand at the bow watching for outcropping corals, and with the outboard tilted so the prop was as high as possible while still in the water, we crossed over into the lagoon without incident. Lowering the engine, I then motored the couple of kilometers to the island, watching for shallow patch reefs all the way. We beached gently on the coral rubble shore, and I listened to the soft slushing as the small waves lapped the shore. That softly sibilant sound, yet much louder than the sound of waves lapping a sandy shore, stays with me, because it sent me off to sleep that evening, and every evening after during my many years doing research at One Tree Reef.

The contrast to Heron Island was surprising. This island was quiet. Heron Island provided a continuous cacophony in summer months courtesy of the muttonbirds that moan all night during the nesting season, a fleet of tractors driving here there and everywhere, and the several diesel generators providing light and power to the research station and the even more power-hungry resort next door. There was no generator, nor were there any tractors, and the rubble island was not attractive to burrow-building muttonbirds. Neither were there any black noddies (although that situation has since changed). These tree-nesting terns build the most amazingly unprofessional nests out of leaves glued together with copious quantities of excrement. On Heron Island, the Pisonia trees are festooned with noddy nests, and especially after rain the air is redolent with eau d'ammonia. Also, after a rain, the noddy nests become extremely slippery as the excrement glue softens, and that's when lots of noddy chicks slip and slide out of their nests and eventually die on the forest floor. Lovely place! On One Tree, during all the years I worked there, the resident terns were of several species, ground-nesting in the rubble, relatively quiet compared to the always cackling noddies, with no hint of ammonia on the breeze. I liked this tiny rugged, silent, fresh-smelling island (fig. 8).

Getting into the water later that day, I was blown away by the lagoon. The large lagoon of Heron Reef is a desolate place scarcely anywhere more than two meters deep, with vast expanses of barren, sandy floor, and only the rare patch reef to provide some structure. There were certainly fish in the lagoon, but not the myriads of species that live wherever there is solid structure and living coral. The main lagoon at One Tree Reef (there are three of them) was rich with patch reefs and long walls of coral, with plenty of open, sandy expanses and lots of structurally more complex habitat as well. And it was in places 6 meters or more deep. Because it was ponded on each low tide but readily accessible from the island at all tide stages, it was also a place where sea conditions seldom interrupted field activities. Not surprisingly,

Figure 8. One Tree Island on my last visit in November 2008. The reef is not at its prettiest at low tide because the relatively high reef flat drains to reveal extensive rubble benches, but that is also the time when a helicopter can land! Photo © Peter Sale.

most research at One Tree Reef has made use of locations in this rich lagoon. I found sites within the lagoon that, because of their exposure, had much of the appearance of a leeward reef slope and many of the species that would occur there. There were other places, much more protected, with delicate coral species not common on outer reef faces.

I had no time to dive the outer slopes on this brief visit, but these also did not disappoint when I got to see them subsequently. The southeastern, more exposed side slopes gently from the reef crest to about 3 meters, then plunges vertically in a 30-meter cliff—a cliff against which it is possible to feel very small if you start imagining the large mouths that might be (almost certainly are) cruising by in the dark blue waters beyond. The northern, somewhat more sheltered side slopes to a similar depth, but more gently. It offers an imposing series of spurs and grooves—spurs projecting out from the reef face and rising 10 to 15 meters above adjacent grooves, and all covered by a densely packed assemblage of coral colonies of many species.

In May 1984, on a visit to the University of Miami, I had my first dive on a Florida reef. Jim Bohnsack took me to Looe Key, then one of just two marine protected areas in the Florida Keys,[3] and I saw Caribbean spur and groove formations for the first time. They were similar in form to those at One Tree Reef but built on a far more modest scale, with spurs scarcely 3 meters above the grooves. Spurs and grooves, of whatever scale, are a feature providing crenulated edges to reefs everywhere, and I suspect that reef geologists have a clear understanding of why and how they form. (I never asked, and they never told me!)

After a good meal, cooked on kerosene-powered stoves, under the hissing light of a kerosene lantern, and a few beers and conversation with Ted and his wife, June, I lay awake on a top bunk, face close to the eaves, listening to the wind swirling gently through the gaps in the rustic hut, the murmurings of some of the birds, and the slushing of waves on the rubble shore no more than 10 meters away. It felt very good, just being there. The following day, we reluctantly bid farewell, and set off back to Heron Island and its semblance of civilization. Once there, I began planning how I might move my field operations to One Tree Island.[4]

This story has said little about coral reefs, but it reveals one facet of their wondrousness: they can be so very different from one another! Heron and One Tree Reefs are of similar age and history, only a few kilometers apart, bathed by the same ocean, and yet both the islands and their reefs are remarkably different, at least in the shallower parts. The richness of One Tree's lagoon is not because it is ponded at low tide. If anything, that makes it less rich because of the less favorable conditions created when ponded water warms in the tropical sun. The richness of One Tree's lagoon arises because it has happened to develop slightly differently. I suspect that geologists may be able to explain why this is so. (The other two lagoons that are better connected to the ocean throughout the tidal cycle are both richer than the main lagoon.) For me, the existence of this difference between these neighboring reefs validates a general impression I have developed over many years: although the patterns and processes leading to reef growth and erosion are predictable, their interactions lead to very different outcomes from one location to another. Whether looking for coral heads of one species, small patch reefs, or larger sites in specific reef habitats, the reef ecologist learns that finding replicate units for use in a study can be damned difficult.[5] Individuals that are very similar to each other, and therefore useful when the experiment will use replicate individuals as controls or to receive the experimental treatment, just don't seem to be available.

♦♦♦

On another occasion later in 1973, I had a single dive on a reef that taught me the same lesson in a very different way. This was the year of the Second International Coral Reef Symposium, a now quadrennial global research conference for coral reef science and management. The Second ICRS was the only one ever held on a cruise ship. Of the approximately 350 passengers on board, 264 were scientists attending the conference. To break even, the organizers—the old and mysterious Great Barrier Reef Committee, which I helped transform into the now vibrant Australian Coral Reef Society about a decade later—were forced to open the cruise up to recreational divers interested in cruising through part of the Great Barrier Reef. We set off from Brisbane on June 22, 1973, sailed up through the Great Barrier Reef lagoon as far as Lizard Island, and then back south, arriving in Brisbane on July 2.[6] Along the way, we attended conference sessions on board and made several stops for field excursions to notable sites.

The MV *Marco Polo* was a clapped-out Russian cruise ship with a surly crew that seemed engaged in some sort of internecine warfare most of the voyage. This was my first ICRS; the thirteenth took place in Honolulu in May 2016 with about 2,500 scientists and reef managers present.[7] On the *Marco Polo* I shared a four-person cabin, deep in the bowels, with Yossi Loya, the Israeli coral ecologist, Professor John Morton of University of Auckland, an older reef biologist toward the end of a distinguished career, and a mad keen diver from Brisbane who woke us all up very early the first morning as he clanked and clunked, putting weights onto a belt and otherwise checking his dive gear in the middle of the cabin's steel floor (I rolled over and went back to sleep).

There were two concurrent sessions of papers given under less than ideal conditions in lounges meant for drinking and partying. Most of us younger scientists spent a fair bit of time in the small bar at the stern and up around the funnels on the top deck exploring chemical influences on brain function. The senior scientists on board, especially those who had organized the conference, appear to have spent lots of time jockeying for position in the nascent hierarchy in our field—years later, some continued to tell tales of the fighting that went on up in the luxury suites! Looking back, as well as some interesting science and the field trips, I remember the friendships that were formed. A friendly rivalry emerged as we Great Barrier Reef natives watched bemused as the Discovery Bay, Jamaica, clique (mostly from Yale University) strutted their stuff. I remember Jeremy Jackson, a tall, afro-sporting white guy about my age, and subsequently a famous reef researcher, complaining frequently that the excursions were not getting him to places where he could dive deep; we believed that there was plenty to fascinate in depths less than 40 meters—this was *not* the Caribbean. It was all a long time ago, but the friendships formed have been renewed regularly at field sites or conferences as we've met again from time to time. We formed a contingent of really old guys who hung about at the Thirteenth ICRS.[8]

At Lizard Island, the ship paused for a field excursion. Some people dived, others snorkeled, and still others went ashore to hike to the peak where Captain James Cook had stood, so long ago, peering out to sea, desperately seeking a passage out of the infernal maze of reefs in which he had got himself trapped.

My Heron Island compatriot Ross Robertson and I had both decided before the voyage that we would forgo the opportunity, and the paperwork involved, to dive. We reasoned that we "knew the GBR" and could enjoy the excursions on snorkel just as well. That decision, which may have demoted us in the eyes of the Discovery Bay clique, meant that we were both among those who had set out for a large patch reef near Lizard Island. (That's the same Ross who would be watching parrotfishes pooping in Panama some years later; in 1973 he was a Ph.D. candidate at University of Queensland who had just astounded the reef science world with his account of the social and sexual behavior of the cleaner wrasse.)[9]

Rolling over the gunwale, I saw the reef face in the distance. A few kicks propelled me toward it, and I dived down for a better look. I came face to face with a giant clam, *Tridacna gigas*, the largest species of clam in the world (fig. 9). It was almost a meter in length and the first one I had ever seen alive. A number of *Tridacna* species occur at Heron and One Tree Reefs, including one named, paradoxically, *Tridacna maxima*. All are substantially smaller than *T. gigas*.

I paused to marvel at its size, and at the vividly fluorescent blues, greens, browns, vermilions, and purples of its mantle, before returning to the surface for a breath. Then I spent a few seconds wondering at the ridiculousness of the old stories of divers drowning because they have inadvertently trodden on a giant clam that has clamped shut around their ankles. *Tridacna gigas* is almost the only *Tridacna* large enough to get your foot into (especially if wearing fins), and this one, like every specimen of this or any other *Tridacna* species I have ever seen, was so plump that its mantle bulged out between its shells, effectively filling all available internal space.

Having disposed of one myth and spent some time watching the clam languidly pump water through its siphon, I turned my attention to the fish and immediately saw a small damselfish of a species I had never seen before. I swam along, mentally ticking off the species I expected, and then saw another species new to me. Ten minutes into my dive, I began to play a game with myself—could I guess the first species I would see around the next corner, and how many new species would I find?

Lizard Island, at latitude 14° South, is substantially closer to the equator than One Tree Reef at 23.5° South, and supports a richer fauna of species of fish, of clams, and of most other taxa. It is a more diverse location, and a number of species present there have ranges that do not extend as far south as One Tree Reef. On the other hand, the One Tree fauna includes a few warm-temperate species whose ranges extend to the southern Great Barrier Reef but not as far north as Lizard Island. But the number of warm-temperate species is quite a bit smaller than the number of tropical species that do not extend south to One Tree Reef, so overall there are more species of fish, of *Tridacna* clams, of corals, of polychaete worms, and so on at this more northerly site.

On my snorkel tour, I was not doing an organized, quantitative survey of

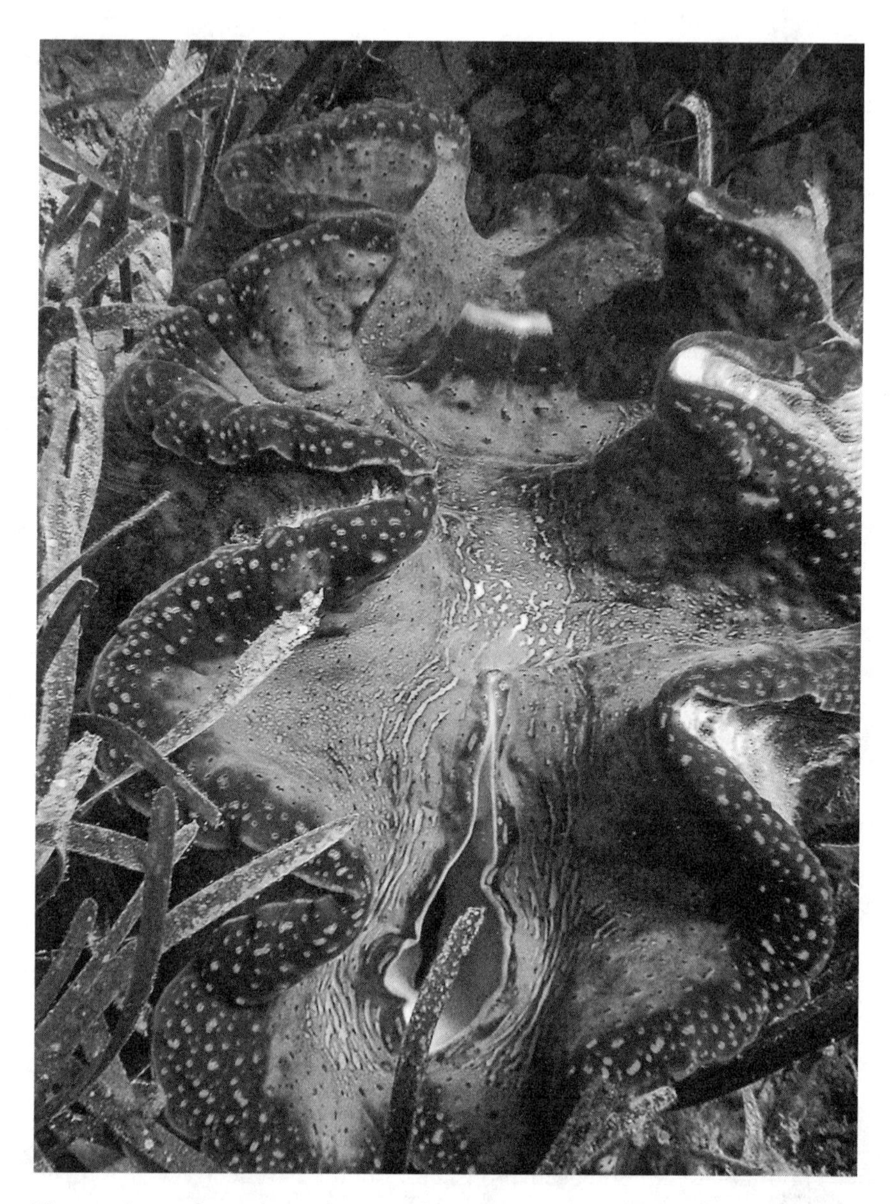

Figure 9. I came face to face with a giant clam, *Tridacna gigas!* This one, photographed off Bolinao, Philippines, in October 2003, is a typical specimen, with a mantle plump enough to bulge out of its shell, incurrent and excurrent siphons quivering slightly as water flows in to be filtered for plankton, and with millions of symbiotic algae in its mantle tissues, being farmed in the tropical sun, and all fluorescing madly in a kaleidoscope of violets, blues, greens, browns, yellows, and even reds. Photo © A. J. Hooten.

the fishes. I was in the water, enjoying myself, exploring a patch reef closer to the equator than any place I had visited until then. But as I continued my game of guessing what I might see next, I became more and more exhilarated by the sheer diversity in front of me. Guessing that next species was really difficult!

This is what people mean when they talk about the richness of tropical ecosystems. It's a general rule across marine, freshwater, and terrestrial environments: head toward the tropics, and the number of species of nearly all taxonomic groups will rise. It's a general rule that ecologists have wondered about for decades—why should this be so? Or perhaps it's a general rule that we have got backward: instead of marveling about and seeking to explain the increased richness of the tropics, we should be pondering the paucity of species in temperate or polar locations. In any event, this pattern was hitting me in the face as I explored a small part of that remarkable reef. I don't remember how many fish species I saw that day, but I returned to the ship bubbling over with enthusiasm for coral reefs. Reefs were remarkable examples of what nature can produce and were also wonderful objects of scientific attention, providing an endless series of questions to stretch my mind. It was a very good day.

♦♦♦

The differences between One Tree and Heron Reefs, which lie right next door to each other, and the differences between One Tree and Lizard Reefs, which are hundreds of kilometers apart, convey the immense variability both among reef architectures and within their ample lists of species. A dive in Mexico taught me just how topographically remarkable reefs can be. This dive took place on the Palancar Reef off southwest Cozumel, just off the east coast of the Yucatán Peninsula. It was 1997, and I was in Mexico exploring possibilities for collaborative research. My Mexican colleagues had taken me to Cozumel to see a project in progress that sought to rescue and transplant corals in danger of being killed by the construction of a cruise ship terminal.[10] The dive was a chance to visit an interesting reef and enjoy ourselves far from harbors and construction.

During the mid-1990s I had realized that collaborating with Caribbean scientists required that I master Spanish. I'd taken some lessons, and this was my first visit with new vocabulary skills in hand. But my colleagues all wanted to speak English, mainly because it was difficult to stop laughing at my hopeless efforts in Spanish. Languages are not my strength. At any event, here I was, in a large, open boat, with a bunch of people I did not know terribly well, who were all speaking rapid colloquial Spanish, while I listened intently, sometimes getting the gist of a few sentences, sometimes completely in the dark.

When we arrived at the dive site, there may have been some discussion of what we would do (about a dozen of us were doing a drift dive, in which the boat would follow behind and pick us up at the end). I learned little of the plan but did not bother to ask because . . . well, it's not rocket science—you roll into the water, descend, enjoy, return to the surface. Also, we were research divers; as a rule, even if we pair up as buddies on the surface, we tend to wander off like so many poorly herded cats once in the water. Anyhow, I rolled over the gunwale and descended to a large sandy arena beneath us, with a wall of reef rising above me to one side. I was one of the first into the water, and floating weightless just above the bottom, I waited for others to arrive. I noticed new arrivals wandering into caves in the reef face, so I swam over and peered into a few caves, too. Nothing particularly special except all were well lit by the sunlight reflecting off the very white sandy floors. Then I came back out onto the arena floor and discovered that nobody else was there. They had disappeared. The boat was gone, too. So, I wandered off across the sandy bottom.

After a minute or so, I finally figured out that I was going in the wrong direction. My Mexican friends had gone into the caves because the dive was a trip through a series of relatively short, but steeply descending caverns that would open in substantially deeper water on the outer face of the reef.

Once I figured this out, I swam into a likely cave and slowly followed twisting, sand-floored channels for a couple of minutes before I, too, emerged on the outer reef face at about 20 meters and saw colleagues scattered in the distance. I remember this dive vividly because the topography was so unexpected. And so vertical. And built entirely of corals, mostly living. I thoroughly enjoyed myself circling tall pinnacles, soaring under arches and overhangs, and going through still more caves and tunnels. I descended briefly to 30 meters, but the reef slope went steeply down well beyond that. Eventually, I ascended the outer face of the reef along with my Mexican friends. The boat was there, waiting, and we got in.

Two other things I remember about that dive—there were hardly any fish, and I used remarkably little air. One of the Mexicans noted that my tank was still half full, in stark contrast to most of the others, once we were back in the boat. I had indeed used less than usual, perhaps because I was so rapt in the splendor of my surroundings that I floated lazily here and there, far more efficiently than my usual effort, and breathing not much at all. As for the lack of fish—I was surprised because caves usually teem with fish, but as I grappled with the linguistic divide I never got a clear explanation, other than that my colleagues could not remember a time at that site when there had been more fish.

FOUR

So Many Ways of Being

It's one thing to talk about the sheer number of species on a coral reef. That can impress people, some more than others, but the richness is far more multidimensional than lists of numbers or names can convey. Amazing biotic richness is a feature of the tropics, and within the tropics, reefs and rain forests rival each other in the bizarre directions in which they take that richness.

The diversity of coral reefs extends across virtually all taxonomic groups yet is far from being fully enumerated. For many invertebrate groups other than the corals, there remain numerous undescribed species even today. Even among the now rather well known fishes, few have common names, and ichthyologists—biologists who study fishes—working on coral reefs continue to describe new species monthly. Ecologists working on coral reefs in the 1960s and 1970s got used to talking about *species a* and *species b,* because even quite common creatures remained undescribed; we also got used to the species for which we used the correct name in one paper becoming known by a different name by the time we wrote a follow-up paper about its ecology. That situation has improved, but taxonomy of reef organisms remains a work in progress—too many species, not enough taxonomists.

♦♦♦

In the early 1970s, a Canadian colleague was in Australia on a sabbatical year and wanting to do a study of the behavior of a suitable reef fish. I suggested that he look at territoriality in the common, easily recognized, shallow-dwelling, territorial damselfish *Abudefduf zonatus*, and he did.

Now, based on specimens collected in New Guinea in 1830, this species had originally been described by the great French ichthyologist Georges Cuvier, who placed it in the genus *Glyphisodon* as *Glyphisodon zonatus*. It's a relatively drab brown little fish with one white saddle about halfway along and two dark eyespots on its dorsal fin. But Cuvier did not know that five years earlier, in 1825, using fish they collected in Guam, French ichthyologists Jean René Constant Quoy and Joseph Paul Gaimard (typically referred to as Quoy & Gaimard) had named the same species *Glyphisodon biocellatus*. Cuvier chose to remember the saddle in his species name, *zonatus*, while Quoy & Gaimard chose to remember the eyespots in their name, *biocellatus*. Subsequent ichthyologists lumped a number of genera of damselfish, including *Glyphisodon*, as belonging to a single genus, *Abudefduf*, making the Western Pacific form *Abudefduf zonatus* and the one in Guam *Abudefduf biocellatus*. In the early 1970s the Heron Island form had been, for more than a century, a well-known, Western Pacific fish called *Abudefduf zonatus*. Seems pretty clear.

My colleague returned to Canada and published his results in a 1972 article called "The Behaviour of *Abudefduf zonatus* (Pisces, Pomacentridae) at Heron Island, Great Barrier Reef." All was correct, for a while.[1]

During the 1970s, considerable ichthyological work was accomplished on the damselfishes by various scientists, especially Gerry Allen (who happened to have shared an office with me while a graduate student of that other great Hawaiian ichthyologist, Jack Randall). Some of this research involved descriptions of more new species in this large family, but a lot involved tidying up earlier decisions about genera. In this process, the rather large (many species) genus *Abudefduf* was recognized as holding a hodgepodge of not particularly closely related critters, and it was split.[2]

Our well-known, easily recognized, shallow-dwelling fish at Heron Island became briefly *Abudefduf biocellatus* (when the equivalence of the New Guinea and Guam specimens was recognized). Then it became *Glyphisodon biocellatus* (when *Abudefduf* was split). There it remained until the mid-1980s, when further research led to further reorganization of damselfish genera and it became *Chrysiptera biocellata.*[3] Its name has not changed since, and I'm pretty sure that this little fish was never particularly concerned about what humans decided to call it.

The same turmoil of name changes was happening to many of the species I was studying in the mid-1970s, and in an article I published in 1979, I decided I had to include a table of synonymies for Australian damselfish species that ecologists had been studying![4] Such is the life of an ecologist working on coral reefs.

On another occasion, I was working with two Sydney University colleagues to explore the rich fauna of what was called resident, or near-reef zooplankton. In the mid-1970s, ecologists had become aware that a substantial proportion of the zooplankton drifting in the water column close to coral reefs was in fact made up of creatures that were part of the coral reef fauna, sheltering in sediment or among interstices in the reef structure during the day but out in midwater feeding at night. The remainder of the zooplankton were open-water creatures washing past the reef. This was a more important distinction than it might seem, because plankton brought to reefs from the open ocean carry nutrients and energy in their bodies, while plankton that have always lived within the reef have bodies containing nutrients and energy that were already part of the reef ecosystem. If substantial proportions of the plankton around reefs were reef-derived that would alter our fundamental understanding of how reef ecosystems obtain the energy and nutrients life needs.

Our first report was a straightforward description of the abundance and taxonomic richness of this near-reef fauna at the Great Barrier Reef sites we studied. We wrote the paper and submitted it to the journal *Marine Biology* for review and publication. In it, we had described a fauna composed mostly of various copepods, plus other tiny crustaceans, plus a range of such other creatures as polychaetes, foraminiferans, and larvae of corals, mollusks, and starfish. But most of these taxa could not be identified and given names—they were undescribed species or unidentifiable larval forms. We included a table of the 114 taxa we had found, some of which probably included several related species. The table was sorted according to families or higher taxonomic groups. In most cases we could confidently assign specimens to genera but not to species. There were a fair number of *species a*, *species b*, and so on.

The editors, as is usual, sent our manuscript for review by other expert scientists, and in due course we received the comments. One reviewer could not believe that we had "not bothered to" identify the species we had collected. The editors were initially going to reject the paper because of this omission. We realized immediately that they had sent the manuscript to reviewers based in Europe and used to working in places with well-known faunas. We had to explain to the editors that there was no way to provide names for the majority of our taxa because they were creatures that had never been described by taxonomists—we were looking at a fauna that was largely unknown. Having to provide that explanation reminded me how different the tropics are to places such as the coastal waters of Europe. Our article was published, without taxonomic names, in 1976.[5]

♦♦♦

Soon after I began making field trips to Heron Island late in 1968, I met Fred and Judy Grassle. As so many biologist spouses do, they had met in graduate school, at Duke University. Fred had pursued his Ph.D. in the lab of Howard Sanders, studying the ecology of benthic organisms in soft-sediment continental shelf locations, while Judy pursued hers in marine invertebrate physiology in the lab of John Costlow. Fred was a member of Sanders's team at the time that they published influential papers on the nature of benthic invertebrate biodiversity and how it varies with depth and latitude (a lot of ecology begins with asking what, where, and how many). On completing their doctorates, Fred and Judy obtained postdoctoral funds to study coral reef systems at the University of Queensland for two years. They were frequently at Heron Island during my visits, and we had many long, memorable conversations concerning ecology, coral reefs, and the organization of life on this planet. Dinner time at the Heron Island Research Station, when Fred and Judy were in residence, was satisfying intellectually as well as gastronomically.

Fred and Judy were working on patterns of distribution of polychaete worms. Most people are not even aware that polychaete worms exist, but they are common in most marine environments and enormously abundant and diverse on coral reefs. The lowly earthworm is technically an annelid worm (actually, "it" is a large group of different species of terrestrial annelid worm), and usually the first annelid anyone meets. At one time, many of us met it academically in high school biology, where it was one of the first organisms we learned to dissect. At that earlier time (when I met earthworms in high school), the relationships among the annelids appeared to be relatively simple. There was a single large phylum (some 9,000 to 17,000 living species) named the Annelida because all its members have a segmented body plan. Annelida included the class Oligochaeta, or earthworms: smooth, cylindrical creatures with tiny bristles on their sides as their only appendages. There were the Hirudinea, or leeches: more or less smooth, approximately cylindrical, but with the nasty habit of sucking blood from host animals, including people. And then there was the enormous variety of almost exclusively marine Polychaeta: segmented worms with a bewildering array of appendages, two per segment. Some polychaetes were sedentary, living within tubes they constructed, while others were mobile, moving about over or through the substratum, much as the earthworms do on land. On this basis, polychaetes were divided into the subclasses Sedentaria and Errantia. Among the sedentary polychaetes were many species of beautiful feather duster worms (family Sabellidae) that protrude flamboyantly colorful, elaborate, featherlike antennae from tubes burrowing deep into massive corals. The worms use the antennae to capture planktonic prey, and if disturbed, they rapidly withdraw them into the tube, appearing to vanish without trace. Among the active species were the fireworms (*Hermodice* sp.), with painfully toxic bristles on their paired segmental appendages, and the palolo worms (*Eunice* sp.). Palolo worms have reduced the eroticism of sex to a minimum—they send the posterior half of their bodies on a one-way trip to the ocean surface, on one night two or three days after the third quarter of the moon in October or November, there to release eggs or sperm, or be captured as a delicacy by Polynesian fishermen. I have no idea if the abrupt severing of the posterior half is a pleasurable experience!

Times change, and the addition of genetic tools has helped change our understanding of the annelids. In place of the Sedentaria and Errantia, the polychaetes are now divided up among some sixteen families, the earthworms and leeches are considered closely enough related to be a single class, and some previously nonannelid worms have been brought within the phylum. Annelid taxonomy and phylogeny (their family trees) are unstable, meaning that the taxonomists are still struggling to sort out this large group of creatures. This is not an unusual situation within the biosphere; we have far less understanding of the relationships among the many different forms of life on this planet than most people realize, and it is only for a few groups, such as birds, butterflies, and perhaps fish, that we can be reasonably confident in identifying any individual found.

The good news is that, despite this taxonomic turmoil, the various groups of species of polychaetes were reasonably clearly separated by differences in their morphology. Identifying individuals found, placing them into specific genera, if not species, is a good enough characterization for a lot of ecology, and it is ecology that the Grassles were investigating. Their interests were not in the phylogenetic relationships among polychaete species but in the patterns of distribution of particular types of polychaetes across reef habitats and the richness (number of species) and abundance (number of individuals of each species) in each habitat. In short, they wondered about the question, How do the many species of polychaetes distribute themselves over the many different habitats on a coral reef?

This may sound like a trivial question, but think about it. Take a more familiar group, such as the birds. In my part of the world (Muskoka region, Ontario, Canada) there are some 148 species of breeding birds and a number of others that pass through while migrating. These 148 range from the osprey and the great blue heron through the common loon and a range of ducks to smaller thrushes, swallows, warblers, wrens, and the ruby-throated hummingbird, plus many others. These 148 species are not distributed evenly across the landscape; each has its particular habitat, its way of life, and its ecological requirements. People who know their birds know where to look, or listen, to find particular species.

Naturally, the birds that occupy a particular habitat are much more likely to interact with each other than with birds that tend to live in quite different places. This sorting of species across the landscape is a general phenomenon and is an important part of how ecological communities are built. Fred and Judy Grassle were interested in how the much larger number of species of polychaetes were distributed across Heron Reef. By discovering this pattern they would add one or two bricks to our growing understanding of how ecological communities are organized. I'm not going to detail all that they found, but I do want to relay one surprising result.

Large numbers of polychaete species live within the rocky structure of the reef. Some of them—important bioeroders—burrow into the dead carbonate skeletons of corals, others get enclosed within cavities formed as the coral grows around them, and still others use crevices created by other burrowing or excavating creatures. One way to sample these secretive polychaetes is to collect samples of living or dead coral and break them into small pieces while continually washing them over a fine mesh screen.

In an article in 1973, Fred Grassle reported on one amazing collection of this type. In his words, "A single head of *Pocillopora damicornis* from 5m depth on the reef slope south of the Research Station weighing 4.7kg (1.6kg living) contained 1,441 polychaete individuals belonging to 103 species."[6] *Pocillopora damicornis* is a common, finely branching, rather beautiful coral; this specimen would have fit within a five-gallon plastic bucket, yet it housed over a thousand cryptic polychaetes ranging in size from a few millimeters to several centimeters in length. They included 921 syllids, 102 capitellids, 68 terebellids, 62 nereids, 56 lumbrinereids, 44 sabellids, and 32 phylodocids. Most of the few larger specimens were eunicids (these are all families of polychaete). As well as these 1,441 worms, the coral head yielded tanaids, amphipods, and isopods (all small crustaceans), as well as decapods (crabs, shrimps), ophiuroids (brittle stars), sipunculids (a type of nonannelid worm), and oligochaetes (back to earthworms). Rereading this paragraph, I imagine a song with mellifluous lyrics—"Syllidae, Oh Syllidae, your Capitellid's waiting . . . ," but I digress.

♦♦♦

The preceding paragraphs contain a lot of technical names for types of creatures most people have never heard of. Why did I write it that way? Because, even today, many creatures that live on coral reefs are poorly known, if at all. They don't have common names and certainly not English names, but if we want to understand the ecology of reefs, we have to recognize that these creatures exist and deal with them. Much of the enormous biotic richness of a coral reef is well hidden as small creatures of many species few of us ever think much about. In the late 1960s, ecologists were still coming to grips with the enormous biodiversity characteristic of coral reefs. The Grassles' work at Heron Island, documenting the distribution and diversity of polychaetes, was part of that process.

At that time, ecologists viewed the world as one in which different species were able to coexist in the environment by having different specializations in habitat, in diet, or in other ecological needs. Otherwise, the theory held, the competitively superior species would outcompete and eliminate lesser species. According to a widely used phrase, each species possessed its own ecological niche. But the tropics (rain forests and reefs in particular) were causing many of us to question that paradigm. Are there really 103 ways to be a polychaete worm that lives within coral limestone (never mind all the nonpolychaete species present)?

The idea that different species had distinct niches was an important part of ecological science through much of the twentieth century. It had its origins in the first decade of the century in the work of Joseph Grinnell and Charles Elton; Grinnell, a plant biologist, emphasized the environmental aspects of the niche, the type of place the species occupied, while Elton, an animal biologist, emphasized the trophic aspects of the niche, the kinds of things the species ate and the other species that in turn ate it. In 1959, G. E. Hutchinson, the great Yale University ecologist, attempted to provide a quantitative definition of the niche as a multidimensional space in which each dimension represented the range of a specific environmental attribute (such as temperature) or requirement (such as food or shelter) that permitted the survivorship of individuals of a species. The volume of such a multidimensional space, or perhaps its average width across selected dimensions, could then be used to define the degree of specialization of that species. A highly specialized species would possess a small or narrow niche, whereas a generalist would occupy a larger or broader niche.

Hutchinson's effort helped ecologists visualize the incredibly woolly concept of the niche and ushered in a decades-long period of theoretical research into such things as niche size and niche overlap among species. He did not help us measure niches, however, because his concept was *n*-dimensional, where *n* was an unspecifiable number, depending on how many different ways the ecological requirements of a species could be visualized. Quantifying some of the dimensions, such as diet, was a similarly unbounded process depending largely on how finely the ecologist chose to characterize the food. Large numbers of creatures got consigned to having narrow food niches because the ecologist did not, or could not, sort out the many types of food they consumed. Lumping many different foods into "detritus" or "plankton" or "grasses" resulted in niches that were far narrower along the food dimension than was really the case. And the same goes for niche dimensions like habitat.

Nevertheless, the idea of the niche, and crude attempts to quantitatively define the niches of particular species, allowed ecologists to see interesting patterns in the groups of species they studied. Coexisting, similar species usually had niches that did not overlap completely, and niche size varied from system to system. Driven by ideas concerning resource limits, competition, predation, and natural selection, ecologists of the mid- to late twentieth century built a seemingly robust body of theory concerning the coexistence of species in nature that now generally goes by the name niche theory. In essence, ecologists expected that species that occurred together in a location would possess different niches, that their niches would not overlap totally and often not at all, and that places so occupied would not usually contain vacant niches (groups of resources apparently not being used). Numerous field studies were done to characterize the way in which coexisting species were distributed across the environment or, for those who could get their heads around multidimensional space, distributed across niche space. It seemed for a while that important increases in understanding were taking place.

Except in the tropics (fig. 10). In tropical systems, or when attention was directed in temperate locations to particularly speciose groups of creatures, these simple rules of species packing did not seem to work so well.[7] Many of us went doggedly on, fitting our observations to the theory (and the imprecision of niche theory provided lots of opportunity to fit the data), but

Figure 10. The sheer number of species is what first hits a visitor to a coral reef. Whether it's the corals, the tiny crustaceans or worms, or the fishes (as in this photo taken in 2006 at Johnston Atoll in the central Pacific Ocean), an attentive visitor finds a lot more variety than would be apparent in other aquatic environments. Mostly, the different species advertise their differences, falling over themselves to present distinct appearances or behaviors. At their best, reefs feed our souls as they reveal the many ways creatures can evolve and how many different species can be present in one place. They baffle us with their richness, and they cause ecologists to wonder if we really understand how ecosystems are put together. Photo © Luiz Rocha, California Academy of Sciences.

some of us started questioning our assumptions. Are there really 103 different ways to be a polychaete worm burrowing through coral limestone? Maybe the rule that no two species could share the same niche was incorrect, even if it was backed up by a robust body of theory that all made logical sense?

♦♦♦

I remember my own awakening to what I'll call the coral reef problem (although it's really a reef, rain forest, African lake, any old diverse community problem). I had arrived at the University of Hawai'i to obtain a Ph.D. working on reef fishes. Before I left Canada, I'd acquired a conventional undergraduate education, including the information then being assembled about how ecological communities were organized. (It was a better education than was occurring in many North American institutions at that time, thanks to my ecology professors who, as well as presenting the conventional niche theory perspective, also drew our attention to the curiously controversial views of two Australians, H. G. Andrewartha and L. C. Birch, and the downright otherworldly viewpoint of Britain's V. C. Wynne-Edwards.) Anyway, despite this education, I left Canada believing that if one could only measure the animal's responses to its environment precisely enough, one would be able to identify the ways in which coexisting species differed and thereby explain how they were able to coexist. I wanted to explain the very high numbers of fish species that occupy most coral reefs. In essence, I wanted to do for reef fishes what the Grassles were attempting to do for the polychaetes. Naturally, I was asked by faculty at the University of Hawai'i what I planned to do for my research, and just as naturally, I told them. Some of them didn't say much; others offered helpful suggestions. But I remember one conversation with Bill Gosline very clearly.

I introduced Bill Gosline in chapter 2 and may have painted him as a bit strange. But he was one of the most respected scientists to pass through the zoology department in the history of that university. His primary research interests were in the phylogeny of the fishes—literally the pattern and process of their evolution into the most varied set of species by far among the vertebrates. He made major contributions at a time when obtaining genetic fingerprints or building molecular phylogenies were science fiction fantasies.

Bill was also a bit of an iconoclast, somewhat irascible, a man who tended to do things his way. By the time I arrived on the scene, Bill was forty-nine years old, growing hard of hearing, and had already decided that the costs for doing his research did not warrant the time and effort required to continue to generate and manage National Science Foundation or other large research grants. Making full use of a coral reef system at his doorstep, and with trivial costs for alcohol, formalin, alizarine, and rotenone and for occasional trips to conferences, he was one of the most productive scientists in his department.[8] He also had a vast knowledge of the coral reef fishes.

One morning, I knocked on the door to Bill's office and went in. Speaking loudly, I told him that I was looking for a useful set of three or four species of fish whose ecological requirements I could compare quantitatively in order to explain how they shared the resources of a Hawaiian reef. Bill listened to my ideas, and then, with only a slight twinkle in his eye, he said, "Well . . . I know just the group of fishes you should study. The moray eels." He paused before elaborating, "Hawaiian reefs include about 30 *puhis;* the eight or so common species of *Gymnothorax,* in particular, are all rather similar. These are about three feet in length, about this big around . . . ." (He held up his two hands, thumbs and forefingers touching, to form a circle.) "They all have sharp, backward-slanted teeth in large mouths; they live in *pukas* in the rock about this big around . . . ." (Holding up his hands again.) "They forage mostly at night. And are piscivorous. They'd be an ideal group for you."

I thought carefully about what he was suggesting, while taking a moment to be impressed with myself for understanding his inclusion of *puhi* and *puka,* Hawaiian for "eel" and "hole." I sensed the considerable difficulty of doing ecological research on such cryptic creatures, never mind such dangerous ones. But it did seem likely that a comparison of some species of *Gymnothorax* would tell an interesting tale.

Then I saw the twinkle, still in his eye, perhaps a little larger. Bill Gosline had just demolished my assumption that Hawaiian reefs would contain fish communities that necessarily divided up niche space according to the prevailing theory. He had told me the tropics had not been reading ecology textbooks! I did not pursue morays further, and while I wasted the first year of my research collecting data on temperature and salinity tolerances of tide-pool-dwelling species—data that might have meant something in the harsher Canadian environment—I put aside the question of how species shared resources on reefs until after I completed my Ph.D. and left Hawai'i for Australia. I've often reflected back to that conversation with Bill. He was far more effective in opening my eyes to what was around me than if he had attempted to confront niche theory directly.

Like most other tropical ecosystems, coral reefs are filled with a superabundance of species of life. Groups of rather similar species, such as Bill Gosline's puhis, are common. Sometimes there are obvious ecological differences when one looks closely at such groups. Sometimes there are much more subtle differences. And occasionally, the differences that exist are so trivial that they are almost certainly ecologically meaningless. This is one of the features that makes coral reefs so marvelous. It is also a feature that can make the scientific study of coral reefs extraordinarily valuable.

♦♦♦

Some years after I had moved to Australia, during the period when I watched territorial damselfishes defending their territories from one another (see chapter 5), I was visiting Stanford University and had a conversation with Joan Roughgarden, one of the United States' leading ecologists, about the ideas I was developing around species coexistence. Joan said something that has stuck with me: "Perhaps we are all asking the wrong question. Instead of worrying about why the tropics are so diverse, we should be asking why the temperate zones have so few species."[9]

Sometimes it is useful to turn the question on its head like that. That conversation with Roughgarden led me, some years later, to develop a seminar trick that I used many times in various venues. In my seminar, I'd begin by noting that the great universities of the world are a temperate phenomenon, and I'd illustrate this with a map of the world in which Oxford, Harvard, and one or two other prominent universities are noted with bright yellow stars and their names. (I always ensured that the university at which I was speaking was included on this map.) Then I'd ask, "What would the science of ecology be like if universities were a tropical phenomenon?" And with a flourish, the stars rearranged themselves so that the major universities were sprinkled across the tropics. I know it's a cheap trick, but that map with the relocated universities got the audience engaged and thinking.

Science develops in university labs. Most ecologists live in the temperate zone and are familiar with temperate ecologies. Some ecologists get to visit the tropics, usually only briefly and only once or a few times.[10] As a consequence, ecologists rarely know the tropics like they know the temperate zone. And temperate ecosystems are less rich in species and governed by a strongly entrained annual cycle in which, for a portion of the year, nothing much ecological is happening. The tropics are not the temperate zone made warmer.

To this day, I believe that if ecology had been a science that developed primarily through studies of tropical systems, many of our ideas about the interactions of species in nature would be quite different to what they are. In places where 103 species of polychaete worm share the insides of a single, small coral head or where a handful of nearly identically sized eel species feed on fish in pukas in the rocks, we might not have developed the idea that species had to be ecologically different in order to occur together. Coral reefs are such places. The diversity of many groups is mind-boggling. But rain forests are equally amazing. This is a feature of the tropics, and if we want to develop a comprehensive theory of ecology for the planet, it has to accommodate the species richness of the tropics.

I am sure that building an understanding of how species are assembled to build ecological communities is a worthwhile goal. Indeed, as we enter the Anthropocene, in which conditions on this planet are going to be changing rapidly, the need to understand the rules that govern the assembly of ecological communities is more important than ever, because we may need to apply a steadying hand to keep ecosystems from devolving into very simple, species-poor, unproductive types. That understanding necessarily requires research into tropical systems, and that research must not be blinkered by ideas developed in much simpler temperate settings. I think that all people will find their lives enriched if they have an opportunity to spend time in the tropics with knowledgeable guides. I know that every ecologist needs to spend time in the tropics as part of his or her education. The natural world is a complex system, and until you see the richness of the tropics, it is difficult to appreciate just how complex our planetary system really is. And coral reefs, of all tropical systems, make that richness so easy to experience.

FIVE

Exuberant Richness

There are a lot of species on a coral reef. In fact, most biologists accept that about a quarter of all marine species—one in four of all species of creature living in the world's oceans—occur on coral reefs. Many of these also occur off coral reefs, but even so, given that reefs cover scarcely 1 percent of the area of the world's oceans, that is a very surprising fact. Still, for me that fact is not what makes reefs so amazing. Telling you that there are lots of species on coral reefs, illustrating this with stories of discovering this richness or of how difficult it is to name them all, still doesn't really convey what these places are like. Because they are not at all like a large, well-organized stamp collection (lots of different stamps, many similar stamps, lots of stamps from each of many nations). It's more interesting and more important to explore the many lifestyles of coral reef creatures. I'll focus on the fishes simply because I know them best.[1]

I used to teach a course on the biology of fishes. I loved that course because it was one of the few courses in our curriculum that exposed students to the diversity of nature, something I value. Over the years, as taxonomists became rare or extinct among university faculty and as the life sciences exploded with so many new topics to teach, the basic richness of life on the planet got less and less attention. This all happened at the same time as students increasingly were entering the life sciences without having a strong background in natural history, based on a childhood and adolescence in the great outdoors. The single Saturday morning field trip to the shore of nearby

Lake Erie was, for many of my undergraduate students at University of Windsor, their first contact with living fishes. The diversity along the shores was an eye-opener: my students did not know most of those species existed. (That was just Lake Erie—imagine if I'd been able to take them to a coral reef!)

My fish course was not a course on fish taxonomy, but I did ensure that students came away with at least a modest appreciation for just how many kinds of fishes swim about. I figured it would be helpful for them to learn about the sheer richness of life, even if for only one portion—the fishes.

I also enjoyed pointing out to those students that using logic, genetic and morphological information, and the rules governing the naming of organisms, it is perfectly appropriate and correct to consider all of the terrestrial vertebrates—the frogs, snakes, turtles, birds, rats, squirrels, hippopotami, dogs, deer, monkeys, and humans—all of us to be, together with the coelacanths and lungfishes, a single subclass of fishes. That's right! The coelacanth and lungfishes, and their fossil relatives, and all the terrestrial vertebrates comprise one group of fishes (the lobe-fins, or subclass Sarcopterygii). There are three other subclasses with modern-day species: the ray-finned fishes (subclass Actinopterygii), the chimaeras (subclass Holocephali), and the sharks, skates, and rays (subclass Elasmobranchii). The lobe-fins and the ray-fins comprise the bony fishes (class Osteichthyes), while the chimaeras and the sharks, skates, and rays comprise the cartilaginous fishes (class Chondrichthyes). I realize as I write this that most readers will have little knowledge of coelacanths or chimaeras and probably could not explain the difference between a skate and a ray, but the wonderful story of the evolution and diversification of the world's fishes is a story for another day—I'm telling stories of the wondrousness of coral reefs, not the wondrousness of all of creation! There are a number of good books that tell about the fishes.

Getting people to understand that we are one species of animal is a first step in building some humility in the face of nature. Getting people to understand that we and all the other terrestrial vertebrates, plus the coelacanth and lungfishes, are one type of fish—that really builds humility, and if there is one thing *Homo sapiens* needs, it is more humility. But I am losing the thread before I start . . . .

♦♦♦

Of all the vertebrate groups, the fishes provide the richest diversity of forms and habits, and the fishes of coral reefs are an outstanding exemplar of fishes more generally. The more we study reef fishes, the more apparent it becomes that there exists an enormous diversity of ways of being a fish on a reef. One might imagine that fish, being relatively large creatures for the most part, would be carnivores preying on smaller fish, crustaceans, and other creatures. That is certainly true for many species. But not for all, and the carnivores reveal a rich diversity themselves. Some of this diversity can be guessed just by looking carefully at their mouths. There are mouths like forceps that pick single zooplankton out of the water and mouths like giant bags capable of opening wide to scoop up plankton by the thousands. There are mouths that pluck single coral polyps one by one and mouths on the end of long, slender snouts that pick god knows what out of deep crevices in the reef. There are mouths armed with flat, crushing teeth for grinding up mollusks and crustaceans and mouths with needle-sharp, backwardly directed teeth to capture agile prey. There are mouths with teeth fused to form a parrot beak and mouths with teeth deep in the esophagus to grind up the food further on the way down. I could go on, but I won't, except to mention the extraordinary mouth of *Epibulus insidiator*, the slingjaw wrasse, common on Australian and Western Pacific reefs and sometimes referred to as "the fastest jaw in the Western Pacific." As its nickname implies, this fish is able to suddenly stick out an otherwise unimpressive snout far in front of its face to capture prey as they flee. According to Mark Westneat of University of Chicago, and Peter Wainwright of University of California Davis, who've analyzed the biomechanics of the system, this wrasse can protrude its jaw by a length equal to 65 percent of normal head length. Protrusion takes only about a thirtieth of a second; acceleration exceeds 100 meters per second squared; and snout speed hits 2.3 meters per second, or over 5 miles per hour.[2] If people ate like that, dinner parties would be dynamic indeed.

Fish also exhibit varied preferences for where they live on a reef, how much space they use in their daily comings and goings, and to what extent they permit others to share this space. Reef fish vary in terms of how social they are and, if so, how complex their social structures are and whether sociality extends to other species. They also vary in how fast they grow, how large they get, how long they live, and whether they fill different ecological roles at distinct stages of their lives. Identifying an animal as a coral reef fish does not really tell us very much about it.

Consider first the simple question of where they live. On a global scale, coral reefs occur throughout the tropics but divide rather cleanly into two major regions containing mostly different species, one for the most part in the Caribbean Basin and nearby locations in the North and South Atlantic, and one across the vast Indo-West Pacific region. Coral reefs of the Galápagos archipelago and western coast of the Americas support a third distinctive Eastern Pacific reef fauna.

The Indo-West Pacific is a reef-sprinkled area of ocean that begins on the east coast of Africa, including the Red Sea, and extends across the Indian Ocean to the west coast of Australia and the giant archipelago that is Southeast Asia. It continues through Southeast Asia, spreading north to Japan, south along the northeast coast of Australia, and on out into the Pacific basin, including all the island groups from the Hawaiian archipelago in the north to New Zealand in the south and to Easter Island in the east. Farther east there is only deep water free of islands and reefs until the Galápagos and the Americas.

A tiny minority of reef fish species occurs throughout the tropics, but the Caribbean and the Indo-West Pacific have been separated for so long, initially by the Eastern Pacific deepwater barrier and a slowly widening Atlantic, and latterly by the rising cordillera of Central America, that faunal evolution has gone on independently in each region and few species are shared. The Eastern Pacific coral reef fauna is a less rich but distinctive mix of creatures, some with Caribbean and others with Indo-Pacific affinities, neither part of the Indo-West Pacific nor part of the Caribbean.

Within the Caribbean region the fauna is remarkably similar from place to place, with many species occurring throughout the region, or through a major part of it, and very few species restricted to locations within it. One major exception is the reefs of northeastern Brazil, which contain relatively more endemic species, likely because the outflow from the Amazon River provides a major barrier to dispersal between these reefs and the Caribbean Basin to the west. Within the (much larger) Indo-West Pacific, the degree of faunal difference is much greater from place to place, yet a large suite of abundant and widely distributed species ensures considerable similarity of the fauna from location to location over hundreds of thousands of kilometers. In the water, looking around, it is easy to tell if you are in the Caribbean or the Indo-Pacific; it's a lot harder to tell where you are within one of these realms, particularly in the Caribbean.

At the local scale, fishes vary in terms of both the breadth of their habitat requirements and which reef habitats they occupy. This is a particularly surprising pattern, because reef fishes almost exclusively produce dispersive larvae that spend a portion of their lives in the open ocean. At the end of that oceanic life—typically a month in duration—the young fishes return to reefs and to the specific habitats that are appropriate for them.

On a typical Indo-Pacific reef, such as One Tree Reef, there is a clear zonation of the physical structure according to depth and distance from the seaward edge, and we accordingly recognize reef habitats such as deep forereef, shallow forereef, reef crest, reef flat, lagoon, and backreef slope. This zonation, initially due to physical forces determining reef growth, is magnified by the fact that so many species of organism occur only within a single habitat or several contiguous such habitats. The pattern holds for sedentary corals and sponges but also for mobile fishes, crustaceans, and mollusks. Animals that could easily travel the 50 meters or so from one zone to an adjacent zone nevertheless remain in their "correct" habitats. This pronounced spatial segregation among habitat zones exists even though the great majority of reef species (not just the fishes) possesses a pelagic larval stage. Somehow, after an early larval life in the open ocean, they get home, not necessarily to where their parents lived but back to the right kind of habitat. How on earth do they do that? (I devote chapter 7 to how they make this incredible journey.) For now, just remember that different places on a coral reef contain quite distinct mixes of species.

♦♦♦

From the first time I put on a mask and took a look at a coral reef, I had the sense that I was visiting a neighborhood. It was not my neighborhood—I definitely did not belong there—but it was filled with creatures going about their daily business. Once I became an ecologist and began more systematic visits, I realized that the creatures I saw so busily engaged were often the same individuals who were at that location the last time I dropped in. And they'd likely be there the next time, too.

There is nothing particularly novel in that observation; it's obvious to anyone who takes the trouble to observe the creatures on a reef. Nor is it unique to coral reefs. The small spatial scale and conspicuous faunal differences among reef neighborhoods are striking, but similar spatial structure exists in a forest full of birds, mammals, reptiles, and insects—just less obviously. Creatures in nature are not scattered randomly across the countryside, and especially not on coral reefs.

What is novel about reefs is the degree to which reef creatures continue their normal round of activities despite the presence of a lumbering diver, usually dark in color with shiny bits, lots of noise, and lots of bubbles. It's not that the organisms are unaware of the arrival of a diver. Make the mistake of arriving, spear or nets in hand, preparing to catch some fish, and every likely target is very well aware of you, suddenly behaving quite differently and far more difficult to catch. The intruding diver is just another large, potentially dangerous predator—to be ignored unless armed and showing signs of hunting.

The sense that a reef location is a neighborhood is likely also enhanced by the fact that reefs mostly occur in clear water. Most of the organisms living there are active during the day, amply using visual signs for communication. While they may also communicate using odors, touch, or sounds, their use of visual signs is one we are well equipped to notice and to understand.

The characteristic that makes a place a real neighborhood, of course, is neighbors—individuals who recognize each other and have established social relationships that affect how they react to each other from day to day. Drop into a village in Thailand or Brazil or Ireland, and at first you will not know much of what is going on. Not knowing the local dialect or language is only part of the problem; you also don't know the existing relationships among the neighbors. Armed with that knowledge of relationships, the cacophony of a marketplace, a pub, or a festive neighborhood barbeque suddenly makes a lot more sense. However, the coral reef neighborhood is socially somewhat more complicated than an Irish pub or a Brazilian street

Figure 11. Dive into a reef such as this one in the Maldives and you enter a neighborhood, a new place filled with creatures going about their day. Mostly, they ignore you unless you seem to be hunting. Photo © Eusebio Saenz de Santamaria of One ocean One breath.

party. The neighbors are of many species, and the interactions among species go somewhat beyond eating and being eaten (fig. 11).

The corals themselves, and other sessile invertebrates, chiefly play the role of architecture in reef neighborhoods. Many reef scientists will take angry exception to that coral-demeaning statement, so I'd better defend it by acknowledging that many mobile creatures are quick to distinguish between living and dead architecture—they care that the corals are alive and well. To extend my defense, let's digress to explore the interactions between corals and certain crabs that live among their branches.

♦♦♦

Crabs of the genus *Trapezia* are small, somewhat flattened, typically crab-shaped crabs found through the Indo-West Pacific region, where they live deep among the branches of living coral. Further, they occur with a specific group of corals, members of the genera *Pocillopora*, *Stylophora*, and *Seriatopora*, all delicately branched members of the family Pocilloporidae, and they don't occupy dead corals. *Trapezia* species, perhaps two centimeters in carapace width, feed on coral mucus and detritus and otherwise act much like typical crabs (from my perspective). However, they are virtually always found as pairs of adults, one male and one female, and usually just one pair of a species per coral head (large coral colonies may harbor pairs of two or more species). Also, they are so belligerent in their defense of their host corals against various coral predators that they have become known as guard crabs. Recent research in Mo'orea has shown that larger species can be quite effective against the large, coral-eating, crown-of-thorns starfish (*Acanthaster planci*), nipping at the delicate tube feet along its arms until it moves away. Smaller species do a good job chasing off the coral-eating snail (*Drupella cornus*) or a smaller starfish, *Culcita novaeguineae*.[3] Corals that have had their crabs removed suffer greater predation—the guard crabs do guard.

The coral gall crabs go a step further in the intimacy of their relationship with their host corals. *Hapalocarcinus marsupialis* is a well-known species. Like the guard crabs, this tiny crab is found throughout the Indo-West Pacific, where it also occurs exclusively with *Pocillopora*, *Stylophora*, and *Seriatopora*, but it looks and acts quite unlike most crabs. Females are mature at a carapace width of just 2.5 millimeters but grow to be over 5 millimeters in width; males are tiny beasts scarcely a millimeter in width. This tiny, quite uncrablike crab gets its name from the fact that females are only found within galls—sites of abnormal growth on the coral branches. Gall formation is induced by the postlarval female crab settling onto a branch tip of the coral. The presence of the crab sitting there disrupts normal growth at that site, and instead the coral grows up and around the female, eventually enclosing her completely in a small chamber except for minute holes through which water can pass. Males are small enough to enter galls, although they get excluded from the most complete galls. Mating by the crabs occurs within galls, but as with all crabs, the females can retain viable sperm for several months. The females live about two years. Like *Trapezia*, gall crabs seem to feed on mucus and detritus, although there is some evidence for filter feeding on plankton.

Gall crabs do not appear to defend their corals and might best be viewed as external parasites in this very lopsided relationship. However, even the relationship between *Trapezia* and its coral host seems one-sided to me. While the name guard crab suggests a two-way relationship, I have to ask, Is the coral engaged in any meaningful way in these social interactions? I'd say no. That the coral benefits from the presence of guard crabs may be ecologically important, but are the guard crabs really doing anything beyond ensuring that their living coral homes continue to live and provide protection and mucus? As a homeowner, I periodically make repairs to the roof or apply a new coat of paint. If my house had living walls, I'd certainly water them and do whatever else might be required. But my house would not become a pet, and certainly not a friend, until it showed some evidence of enjoying my presence. Corals don't enjoy their crabs, so far as I can tell.

Similarly, many hundreds of species of fishes, chiefly gobies, live among the branches of living coral, within the tubes of sponges, or up the anus of sea cucumbers. In all these cases, the hosts must be alive, but they certainly do not have to behave like partners in a social relationship. So, I stand by my claim. So far as I know, the corals, sponges, and other sessile creatures of a reef neighborhood are primarily playing an architectural role. The inhabitants of these neighborhoods are the fish, crabs, snails, starfishes, and other mobile creatures, and they interact socially with one another.

♦♦♦

There are thousands of species of shrimp on a reef, and many of them live in specific relationships with other sessile or mobile creatures. Lots of these relationships are of the pearlfish–sea cucumber, living-up-the-anus variety, with not a great deal of social behavior taking place.[4] Living on the surface, between the spines, or in the gill cavity of a much larger creature, as some shrimps do, is not a signal, by itself, that shrimp and host have a social relationship. But then there is the partnership between burrowing shrimps of the genus *Alpheus* and fishes of several genera of goby.

There are many hundreds of species of alpheid shrimps (family Alpheidae), called snapping or pistol shrimps because of their ability to create loud snapping sounds using their larger claw. Many alpheid shrimps live on coral reefs. Their snaps are one of the dominant sounds in reef waters. *Alpheus* is one of the most speciose genera, with 280 species named and possibly 1,000 species when the not-yet-named ones are included. Many of these shrimps live symbiotically with other creatures—corals, sponges, starfishes, sea urchins, mollusks—while others burrow into the substratum. And many of the burrowing species share their burrows with a range of goby species (fig. 12). In these partnerships one or a pair of shrimps build and maintain a burrow, and one or (rarely) a pair of gobies share the burrow with them. The shrimp maintains the burrow, both shrimp and goby use the burrow for shelter (and for breeding), and the goby signals approaching danger to the shrimp, which happens to be nearly blind. Reportedly, gobies also sometimes move pieces of algae into the burrow, which the shrimp (but not the goby) then eats. The goby spends most of its time just outside the mouth of the burrow, and the shrimp sits at the burrow mouth, its antennae on or near the back of the fish. The goby signals the approach of predators by very specific fin flicks that are communicated by touch to the shrimp. A goby deprived of its shrimp does not perform these signals. Closely related species of goby build their own burrows, but the ones that cohabit with shrimps never build burrows of their own, even if deprived of shrimps.

The relationship between the burrowing shrimp and the goby is a type of mutualism clearly beneficial to both partners. The most interesting thing about it, from my perspective, is that even though this develops as a response between one goby and one shrimp and involves sophisticated communication between the partners, it is not a strict species-to-species partnership.

It is common, throughout the Indo-West Pacific, to find six to eight species of *Alpheus* occurring with a similar number of species of goby, and while there may be preferences shown for particular species as partners, individuals of each goby species occur with individuals of several, or all, of the

Figure 12. Several small gobies share burrows with shrimps of the genus *Alpheus*. This goby is *Amblyeleotris steinitzi*, photographed in Palau, December 2016. Note its size relative to the piece of *Acropora* branching coral rubble just above and to the right of the fish. The shrimp is at the burrow mouth with its tail in the entrance. One of its antennae is draped over the back of the goby, touching the edge of its dorsal fin. The goby is in an alert, heads-up posture. Photo © Luiz Rocha, California Academy of Sciences.

shrimp species and vice versa. For example, Ilan Karplus, of the Hebrew University in Jerusalem, who has devoted much of his research career to studying these associations, reported on a study with two colleagues in 1981 that looked at the specificity of the partnership.[5] Their work was done at the tiny Eilat Coral Beach Nature Reserve on the shores of the Gulf of Aqaba, northern Red Sea.

Six of the nine species of goby and seven of the eight species of burrowing alpheid shrimp known to occur in the northern Red Sea were found at their study location. Four shrimp species were restricted to lagoonal sites less than 2 meters deep, where they were spatially segregated, likely because of differing preferences for sediment type (which varied with distance from the reef face). The other three species were restricted to deeper reef slope sites but were intermixed in depths down to 20 meters. The most common goby, *Amblyeleotris steinitzi*, occurred at lagoonal and reef slope sites, and shared burrows with all but one of the shrimp species. *Ctenogobiops maculosus*, the second most common goby, also occurred in both lagoonal and reef slope locations and shared burrows with all seven shrimp species. *Eilatia latruncularia* was rare (only four seen) and restricted to reef slope sites, where it occurred with two of the three shrimp species. The remaining three goby species, *Cryptocentrus cryptocentrus* and *lutheri*, and *Lotilia graciliosa*, were restricted to lagoonal sites where the two *Cryptocentrus* species shared burrows with the same two shrimp species, and *Lotilia* occurred only with one of these.[6]

Despite the tendency for gobies to pair up with several different shrimp species, each showed a preference: *Amblyeleotris* for *Alpheus purpurilenticularis*, *Cr. lutheri* and *Cr. cryptocentrus* for *A. djiboutensis*, *Ctenogobiops* for *A. rapax*, and *Eilatia* for *A. rubromaculatus*. (*Lotilia* was too rare to test.) Similarly, every lagoonal shrimp species was most commonly seen with one goby species (four different gobies), but the reef slope species of shrimp did not show preferences for particular gobies. This study is particularly well done, using an analysis that accounted for habitat preferences of gobies and shrimp first and then asked whether there were also preferences for burrow partners. (A particular species of goby could occur with a particular species of shrimp simply because they both preferred a particular habitat.) Other studies in other regions have told an essentially similar story—there are clear partner preferences, but both shrimps and gobies pair up with multiple partners.

One can dismiss these partnerships as simply cases of two types of animal having evolved behaviors that permit them to share a burrow. But think for a minute. The juvenile goby (or shrimp for that matter) has a larval period out in the plankton. It then returns to the reef and knows enough to seek out an individual of this radically different type of animal to share a burrow with.

Even if the behavior is strongly genetically controlled, it's still amazing. There isn't yet any evidence that individual partners recognize each other, although once the partnership is formed, it can continue so long as both are willing to share a burrow, so I'd not be surprised to learn that they do. What about more complicated neighborly relationships?

♦♦♦

Teaming up to accomplish goals is a common occurrence among reef fishes. Multispecies schools of grazing herbivores move across a reef together, and smaller, mixed-species groups of benthic carnivores forage across the bottom. Pugnaciously territorial damselfish herbivores tend to cluster in tight, sometimes mixed-species groups, even though this keeps them perpetually busy defending their borders from each other.

The mixed-species schools of grazers are more like herds of cattle than a school of sardines, herring, or tuna; the individuals do not coordinate their swimming into a tight ballet, the group moving as one. Instead they wander over the reef together, grazing as they go. Such groups typically include species of parrotfishes, surgeonfishes, and, in the Indo-West Pacific, rabbitfishes (Siganidae). There may even be a few benthic-feeding carnivores such as goatfishes (Mullidae) or wrasses (Labridae) mixed in. Such schools are impermanent, and individuals are continually joining and leaving as the group enters and then leaves a region of reef that is home to each. Though reasonably common, mixed-species schools are not always present on every stretch of coral reef. Nevertheless, schools can number in the thousands of individuals and are a potent force mowing down algal turfs across the reef whenever they form (fig. 13).

Individuals within the group sometimes appear to be traveling more closely with conspecifics than with other species present, but I don't believe anyone has ever tested whether there is a real preference for joining a group of one's own rather than another species. It seems to me that participants recognize that the other members of the group are engaged in essentially the same behavior, regardless of species. Whether the odd goatfish in the school understands that she is hunting for small invertebrates in the sediments of the reef while the parrotfishes beside her are grazing algae I do not know.

These herbivores travel in large feeding schools for a good reason. Much

Figure 13. Multispecies schools of herbivores are common on reefs, and individuals join and leave a school as it passes by the part of the reef where they live. This small school off Kailua Kona, Hawai'i, chiefly of manini, *Acanthurus triostegus*, and brown surgeons, *Acanthurus nigrofuscus*, includes one yellow goatfish, *Mulloidichthys vanicolensis*, plus a couple of other species lurking behind. Photo © Kris Bruland.

of the reef surface suitable for growth of algae is being actively defended by clusters of territorial damselfishes, surgeonfishes, or parrotfishes.[7] Individual roving herbivores (or benthic carnivores) have little chance to penetrate the well-defended territories, but a few hundred such fish can swamp the defenses of the group of territory holders, permitting at least some grazing within territorial borders before being chased out. The social interactions involved do not require recognition of individual school members (although I'll bet some of that happens), but there is clearly recognition by each individual that joining the school for a few minutes or hours is the best way to feed effectively. In all likelihood this is learned behavior, raising the possibility of local or regional cultural traditions.

The behavior of territorial damselfishes has been well studied over many years. These mostly rather drab, 10-to-12-centimeter-long fishes remind me of bulldogs or testosterone-driven gang members with permanent chips on their shoulders. Looking at a group of territory holders, my first impression is of frenetic pugnacity, with every individual seemingly on high alert, perpetually guarding its borders (the territories are typically less than a meter wide) from its neighbors and attacking every other creature in sight that looks as if it might be about to invade. I spent many hours in the early 1970s watching such groups, so I know that this first impression is mistaken (it happens partly because you—big, black, lumbering, noisy, bubble-spewing—have just arrived on the scene). There can be periods of bucolic calm, as fish tend the algal turf within their territories, alternately feeding and pruning what have been called their gardens, removing detritus that might get washed in, encouraging sea urchins to move elsewhere, all while occasionally looking up to check on their neighbors. These calm periods are interspersed with times of much more aggressive action, directed at neighbors and intruders, but even this activity is carefully targeted.

Most people first learn about territoriality in birds, where territorial defense is usually directed toward other members of one's own species. That is seldom the case on the reef. Territorial damselfishes defend their territories from a broad range of other species. But the defense is not haphazard. Species most likely to browse on or disrupt the animal's algal turf garden, or, for breeding males, species most likely to feed on eggs in the nest, are attacked. Other species are ignored.

How do we know about this fine discrimination? Credit goes to the late fish behaviorist Arthur Myrberg of the University of Miami, and particularly the Ph.D. work in the early 1970s by his student Ron Thresher, now retired and angling for trout in Tasmania. Myrberg called the common Caribbean threespot damselfish (*Stegastes planifrons*) "pound for pound the most dangerous fish in the sea." Ron Thresher used "model fish" experiments in the field to measure the likelihood of a territorial three-spot damsel attacking an intruder.[8] Ron's models were living fish of various species, enclosed in clear plastic containers that could be positioned at specified distances from the center of the test individual's territory. He reasoned that the distance the territory holder would go to attack was a measure of the intensity of defense against that species. He showed a surprisingly reliable pattern of differences among test species that made perfect sense if you examined their feeding or other characteristics. Subsequent work by others has shown this to be a general characteristic among territorial damselfishes—territorial defense is directed to a large number of species that might disrupt the algal food or steal eggs from a nest, but not to numerous other species. Such observations tell us that damselfishes are quite capable of telling one species from another and learning which need to be attacked.

Early in my time in Australia, I became interested in territorial damselfishes because I noticed that groups of territory holders were frequently made up of several species. Each defended its territory from its own and from the other species present in the group, as well as from all those transient herbivores and egg predators. I wondered how it was possible that several species of damselfish could be ecologically so similar that they would form mixed-species groups of territory holders packed tightly together.

My approach was to spend many hours, spaced over several years, watching territorial damselfishes. I've learned to recognize individual damselfish that I found defending the same, or similar, boundaries year after year. I've seen newly settled juvenile damselfishes 1 centimeter long successfully beat off the attacks of 12-centimeter-long neighbors and gradually expand their territories at the expense of those neighbors as they grew up. I've seen territories of fish that had disappeared be divided up by neighbors. I drew maps and recorded changes in the boundaries of the territories over time, expecting to find fine-scale differences in habitat preferences among species. I expected that the mixed-species groups were actually intermingled groups, but with each species occupying interspersed patches of particular, subtly different types of habitat. I also did some simple removal experiments to see what happened when territorial space was liberated. I described my research as monitoring the real estate transactions among damselfishes.[9]

The more data I collected, the more convinced I became that the mixed-species groups I was watching were in fact groups of species that used the same type of space on the reef. These fish were competing for space, and space occupied by one species could, over time, become occupied by other species. This led me to thinking about larval dispersal and subsequent settlement to coral reef habitat as an essential part of the explanation for how similar species of fish coexist on reefs. I called it lottery competition, and because this idea contradicted the prevailing paradigm that each species in a community must possess a unique niche for coexistence to be possible, my work attracted a lot of attention and a fair bit of argument. (I'll take this story further in chapter 6.)

Given that territorial damselfishes (in contrast to the many nonterritorial, brilliantly colored, far more attractive ones) are territorial throughout their postlarval lives, we might wonder how they get to know much about the social group to which they belong. Each has neighbors with whom he or she is periodically fighting, and frequent intruders of various species to be dealt with, but each is stuck within a territory rarely 1 square meter in area—a bit like living in a high-rise apartment but never opening your door except to shout at a neighbor.

Watch a group of territory holders for a few hours, however, and you ultimately see a third kind of activity—a sudden cessation of hostilities among neighbors and a rapid coming together as a group that then moves quickly through the territories of each individual in turn before, just as rapidly, ceasing, returning home, and resuming belligerent hostility. This is a nearly impossible behavior pattern to study because it occurs so rapidly, so rarely, and so unexpectedly. My sense, after seeing it a few times, was that it could occur at any moment but was most likely during the summer breeding season.[10] I think that although they may be belligerent beasts, they have ways of interacting positively that are doubtless essential if any reproduction is ever to occur. This visiting seems to be a way of finding out who the neighbors are and where the prospective mates live—a cocktail party for belligerent neighbors.

Goatfishes or wrasses, feeding mostly on benthic invertebrates, commonly join herbivore feeding schools and catch prey disturbed by the grazing herd. Small groupers, trumpetfishes, and other piscivores also join schools of herbivores, using the parrotfishes as a moving blind within which they hide while awaiting opportunities to strike at prey—likely some agitated damselfish trying desperately to defend its territory from all those parrotfishes. There are also much smaller mixed-species gangs of predators roving about on a reef independently of any herd of herbivores. These foraging groups are similarly transient, with individuals joining and leaving, but they rarely include more than three to six individuals at any time.

Steve Strand provided one of the first detailed accounts of the behavior of such groups in 1988, based on observations made on a rocky reef in the Gulf of California, yet his description also fits what is seen on coral reefs.[11] He distinguished "nuclear" and "attendant" species. The nuclear species were typically larger and occurred as single individuals, whereas the attendant species were smaller, usually present as a small group of one to ten individuals of one or more species. He also distinguished between primary and secondary nuclear species. At Strand's site, a common moray eel and an octopus were the primary nuclear species, followed by attendants whenever moving over the reef, while nineteen other fish species including parrotfishes, triggerfishes, rays, and grunts were followed frequently to occasionally by attendant species, but only when actively foraging. The attendant species included small groupers, wrasses, porgies, and other benthic-feeding carnivores. All nuclear species fed in ways that disturbed the substratum, and a primary benefit for all attendant species was access to food organisms flushed or displaced by the feeding of the nuclear member of the group. As in mixed-species herbivore schools, individuals continuously join and leave such groups, remaining members for periods of minutes, seldom as much as an hour.

That individual fish join up and follow only certain species, and that most followed species are only followed while they are actively foraging, tells me that this is learned behavior by attendant species and tolerated by the nuclear species. Whether individuals of particular attendant species differ in their willingness to join up with nuclear fish or have individual preferences among nuclear species is not known. Nor do we know if individuals of attendant species recognize individuals of nuclear species or even come to time their own activities so that they are at the right place and time to take advantage of individual nuclear species that come by. Nor do we know if reef herbivores time their activities to be "waiting at the curb" when the herbivore school passes by. If such detailed coordination among individuals of different species of fish were discovered, it would not surprise me—neighbors on coral reefs know and interact with each other. Indeed, I remember vividly the regular afternoon appearances of an old green turtle missing one rear flipper that would float languidly past one site where I watched damselfish territorial defense. Old and slow, this turtle carried an unusually dense algal turf on its shell. As it floated into view on the afternoon tide, my damselfishes ceased their border patrolling to rise up the couple of meters to feast briefly on this algal buffet. I thought of that turtle as the local ice cream truck, and maybe the damselfishes did, too.

♦♦♦

Another feeding association by carnivores that is widespread on reefs has recently been looked at with surprising results (fig. 14). From time to time, a moray eel and a grouper team up to hunt.[12] To the casual diver, the grouper appears to be following the eel as it hunts. Strand reported it this way, but research has since shown it to be a more complex and equal partnership. That research emerged from the laboratory of Redouan Bshary, who began his career in primate behavior in 1995 with a Ph.D. from the Max Planck Institute for Behavioral Physiology in Seewiesen, Germany, and with field studies in Africa. A few years later, visiting the Red Sea, he fell in love with reefs and reef fishes. And in 1998 he first saw what he took to be signs that a grouper was signaling to a moray, "Let's go hunting." About a decade later, Bshary published his first paper on groupers and morays. By then, the notion that groupers often teamed up to hunt with morays, with octopus, or with larger wrasses had been nurtured by observations across the Indo-West Pacific and in the Caribbean—but nobody had looked closely at the behavior involved.

Redouan Bshary discovered that this behavior involved a lot more than fish casually following each other about.[13] He showed that groupers and eels would occur together more often, and remain together far longer, than might be expected if they moved independently around the reef. Pairs remained together for up to 93 minutes when the expected time together (based on the numbers of eels and groupers in the area) was less than 2 minutes. During this time, they also remained within one to three grouper body lengths, interacting with each other. In addition, the groupers used a specific visual behavior to signal to the morays. It was a rapid, repeated head shake (three to six shakes per second) shimmying movement, with the anterior portion

Figure 14. Cooperative hunting by reef predators. The top panel shows a roving coralgrouper, *Plectropomus pessuliferus*, hunting with a giant moray, *Gymnothorax javanicus*, in the Red Sea. In the middle are a spotted moray, *Gymnothorax meleagris*, accompanied by a blue goatfish, *Parupeneus cyclostomus*, and a young blue ʻulua, *Caranx melampygus*, off the Kona coast of Hawaiʻi. The bottom panel shows a coral trout, *Plectropomus leopardus*, in Alex Vail's experimental tank with the cutout moray model. Top and bottom images © Alex Vail; middle image © Kris Bruland.

of the dorsal fin kept lowered but flicked up and down very rapidly (0.1–0.3 seconds per flick) while the head moved side to side. Morays responded to this behavior by emerging from their shelters and moving with the grouper. A modified form of head shake, in which the grouper positioned itself head-down over a hidden prey organism, was used to point out hidden prey to the partner. Otherwise head shakes were not seen. Bshary also suggested the head-shake behavior would not occur if the grouper had recently eaten; it was a signal meaning, "I'm hungry, let's hunt." Although the morays responded to the signals from the grouper, Bshary did not see any signs that the morays signaled back. He also showed that, although the fishes did not share their captures, hunting was more successful for both fish when working together.

The story of the grouper-moray partnership does not end there. Research done at Lizard Island by Alex Vail, a Cambridge Ph.D. student mentored by Andrea Manica and Redouan Bshary, led to intriguing papers in 2013 and 2014. The first article, in *Nature Communications*, showed the signaling by Red Sea and Australian groupers to be a surprisingly sophisticated referential gesture—the grouper is telling the moray to look at, or pay attention to, some other object or location. This was something that many psychologists would not expect in animals other than primates. The 2014 paper in *Current Biology* took analysis even further.[14] This paper described a novel set of aquarium experiments run in Australia and using the coral trout, *Plectropomus leopardus*, as the grouper subject. This species, like groupers elsewhere, sometimes hunts with morays, large wrasses, and other benthic-feeding carnivores including octopuses, and uses the same signals as do Red Sea groupers.

Alex Vail had the luck to grow up on Lizard Island, where his parents are the codirectors of the Lizard Island Research Station, perhaps the best-run field research facility on the Great Barrier Reef over its forty-year history. Lizard Island, offshore from Cooktown, is a high island, a continental rather than reef-built isle. In July 1770, while waiting for his ship, *Endeavour*, to be repaired on the beach at the mouth of the Endeavour River, Captain James Cook went by small boat out to Lizard Island and climbed to its highest peak.[15] From there, he discerned a route out through the maze of reefs, beyond the outer barrier, and into the Coral Sea. In one sense, Lizard Island made it possible for Cook to complete his first circumnavigation. The Lizard Island Research Station, which shares the island with a high-end resort, was established by Frank Talbot in the mid-1970s and is operated by the Australian Museum.

The coral trout is a common grouper on Australian reefs. Alex set out to determine experimentally whether the coral trout makes rational decisions about whether to solicit the help of a moray eel when foraging. He used a technique as old as the study of animal behavior, the use of model animals. The first use of models to test the behavior of a fish, to my knowledge, occurred in the late 1930s when Nikolaas Tinbergen undertook his classic studies of social behavior in the three-spined stickleback, a tiny fish common in small streams and ponds in his native Holland. Building simple models fastened to wires that allowed them to be manipulated in the aquarium occupied by a male stickleback, he showed that a model could elicit courtship behavior if it had a swollen belly mimicking a female with eggs or aggressive behavior if it had a red belly like another male. Beyond that, the model could be quite crude, scarcely resembling a stickleback at all. Since that time, simple models have proved useful in many studies of fish behavior. (Ron Thresher, mentioned earlier, could have used models in his studies of damselfishes, but he chose instead to use living fish in clear containers.)

Whereas Tinbergen used sticklebacks about 5 centimeters long, housed in small aquaria, Alex Vail worked with coral trout 46 to 59 centimeters long. All test animals were collected from the nearby reefs and housed for a few days in large tanks before testing. For the experiments, they were placed singly into a tank about 2 meters in diameter and 50 centimeters deep that contained two structures—a moray hiding place to one side and a prey shelter at the other side of the tank. The prey shelter was completely surrounded by a clear plastic cylinder extending above the water surface so that the trout could not actually catch the prey. The prey, a dead frozen pilchard on a wire, could be wiggled about in a crevice beneath its shelter, outside the crevice and at the base of the shelter, or in the water column above the shelter. The moray eel was a life-size, two-sided, laminated photograph cut to shape and weighted to sit on the bottom of the tank. This was rigged on fishing line so that it could be moved backward into or forward out of its shelter and toward the prey. By moving the moray toward the prey while simultaneously moving the prey up to the top of its shelter, Alex could simulate the flushing out of the prey by the eel.

Experimental trials began with the moray moving 20 centimeters out of its shelter. Then the prey was presented at the base of its shelter and wiggled until the coral trout saw it. Next, either the prey was pulled immediately to the water column above its shelter, or it was pulled to the center of its shelter, within the crevice. In the first case, the trout should ignore the moray and proceed to attempt to catch the clearly visible prey. In the second, it should solicit the help of the moray to flush out the prey.

Alex asked the question, "Will the coral trout solicit the help of the moray only if the prey is inaccessible in its crevice?" He answered that not only would it do that, but it did so by using the same signals reported for groupers in the field, and it was at least as proficient as a chimpanzee in making that decision (fig. 15)! (An earlier study by others had examined the ability of a chimpanzee to recruit a second chimp to help solve a difficult task and obtain food.)

I've gone into detail about this study partly because the idea of a fish of one species assessing a problem and seeking help from a second species is probably new to most people. But I also want to emphasize how simple experiments can reveal so much about what animals do and how they do it. In the real situation, there is a second species involved: the moray responds to the invitation and joins the hunt. We do not yet know whether all species of grouper and of moray communicate this way. Neither do we know, for the species that do hunt together, whether every individual uses this hunting technique. Nor do we know whether groupers must learn the correct signal by seeing other groupers use it or somehow know it instinctively. Nor whether morays must learn to interpret this signal. What we do know is that in some places on coral reefs, morays and groupers of some species hunt together, making apparently rational decisions about when to ask for help and how to respond. And we also know that some groupers and morays (and let's not forget some wrasses and octopuses) recognize each other as living creatures with an interest in catching similar prey, but with particular hunting skills most beneficial under specific circumstances. The species in coral

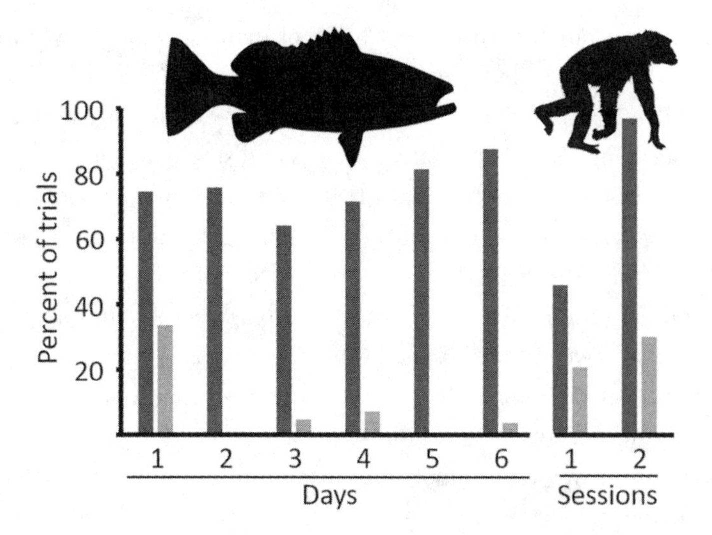

Figure 15. The results of Alex Vail's experiment are compared to results for comparable experiments using chimpanzees. On each of six days of testing, the percentage of trials in which the coral trout signaled to the moray model is plotted for (1) those trials where signaling for help was appropriate (dark bars), and (2) those trials where signaling for help was inappropriate (light bars). After the first day, the fishes seldom behaved inappropriately. The chimpanzee tests, by other scientists, required a subject chimp to signal a second chimp to help in obtaining a food reward when such help was necessary (dark bars) and when inappropriate (light bars). The two sessions were "a few days" apart. Coral trout actually did better than chimps!

reef neighborhoods are engaged in far more complex social interactions than if they just swam about looking colorful for tourists.

♦♦♦

Everybody of a certain vintage remembers that cantina on Tatooine, a strange, dark, smoky place where all sorts of people gathered to imbibe their favorite libations, listen to the band, talk to one another, perhaps pick partners or fights. It's where Luke Skywalker first met Han Solo, and is definitely out of this world. In many ways, a cleaning station on a Pacific coral reef is the Mos Eisley Cantina in real life. It's not frequented by extraterrestrials, and it functions more as a spa than a bar, but it brings the diversity of reef species together in the most complicated interspecific sociality to be found anywhere (fig. 16).

A cleaning station is where reef fishes of many species go to have their ectoparasites removed by other species that have specialized to do this cleaning activity. Fish, in one sense, are simply mucus-covered surfaces ideal for a small army of parasitic trematode worms and small crustaceans, particularly many isopods, to nibble on. Fish lack arms or similar appendages with which they can scrub their own backs, let alone get in among their teeth or gill arches and floss regularly. Sometimes a fish will swim down and glance off a sandy substratum, using the sand to scrub its sides, but this misses those parasites that have got into places at the base of fins or in among the gills. Allowing other creatures to come aboard and remove these undoubtedly itchy nuisances seems a logical way to behave.

There are many ways to look at a cleaning interaction. Ever since 1961, when diving pioneer and marine biologist Conrad Limbaugh described cleaning symbiosis in *Scientific American*, this phenomenon has been one of those standard stories whenever life on the coral reef is being described.[16] That his article was filled with wonderful underwater photos (rare at that time) no doubt added to its effectiveness. Limbaugh's report was based on skin diving in the cool water off southern California, but although cleaning symbiosis is widespread in marine and freshwater fish communities, it is just much more conspicuous, more complex, and more downright amazing on a coral reef. Nearly all the research investigating aspects of this phenomenon has been done on reefs.

Think first about cleaning as a rather strange, shared behavior between two partners. Cleaning, the removal of external parasites and occasional bits

---

Figure 16. Cleaners are readily seen plying their trade on reefs. Here (top) a cleaner wrasse, *Labroides dimidiatus*, emerges from the mouth of a spotted sweetlips, *Plectorhinchus pictus*, on the Great Barrier Reef, and (bottom) a group of cleaning gobies, *Elacatinus* sp., have set up shop on a Bahamian coral head and one is checking out the creases behind the lip of a tiger grouper, *Mycteroperca tigris*. The center panel shows that cleaner wrasses, here *Labroides phthirophagus*, will even hunt for food along the sleek body of a needlefish, *Platybelone* sp., or on manini, *Acanthurus triostegus*, not much larger than themselves. Two of the manini are posing with one gill cover raised, inviting the cleaner to poke around among the gills. Upper photo © Luiz Rocha, California Academy of Sciences; bottom photo © Isabelle M. Côté; both center photos, taken on the Kona coast of Hawai'i, are © Kris Bruland.

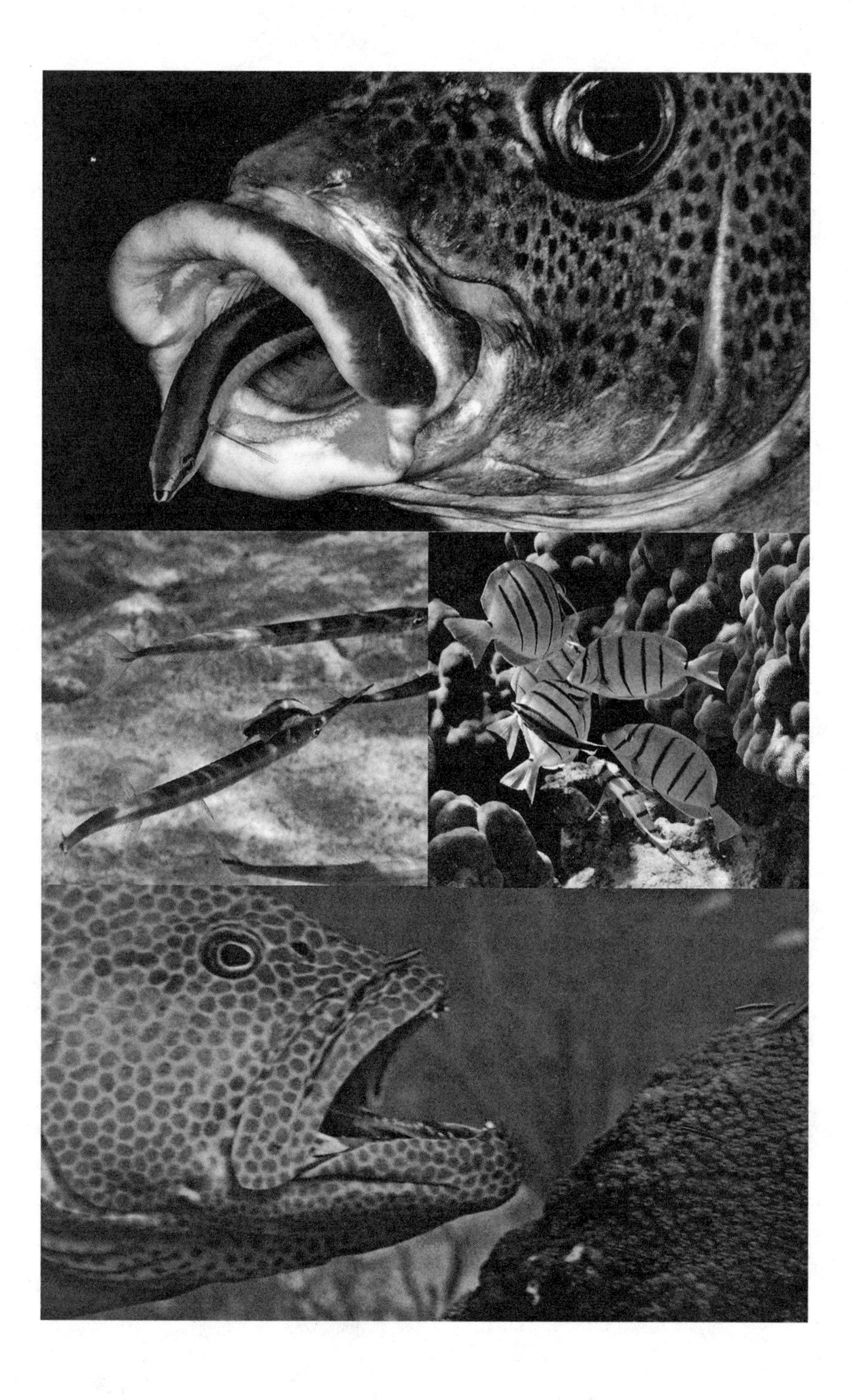

of flesh from the surface of a (usually larger) host, client, or "cleanee" fish, is a behavior practiced by a number of small reef fishes and reef crustaceans. All cleaners share a bizarre willingness to come very close to larger, often predatory species, usually while "dancing" conspicuously. They do this to make contact and then to move about over the client's surface, and into its mouth, throat, and gill chambers, while consuming minute ectoparasites of various types.[17] For most cleaners this risky activity is an intermittent or a transitional pattern of behavior used during juvenile life. In the Caribbean, for example, a number of smaller wrasses are known to clean other fish while young but cease this behavior as adults. Wrasses of the Indo-West Pacific genus *Labroides*, however, clean throughout life as their primary way of making a living. During a day, a single *Labroides* might have over 2,000 individual cleaning bouts, with clients belonging to perhaps 150 species (some clients will visit cleaning stations repeatedly during the day, being cleaned as many as 144 times). For any cleaner, every decision to clean is fraught. Small fishes are surrounded by larger fishes with mouths big enough and often with an interest in eating little fishes. Cleaning requires that the little fish risk a sudden demise by putting him or herself close to those large mouths. The small shrimps that clean also face this dilemma, and if they ponder this decision, they face the same challenge as the small cleaner fishes: Is the chance to forage over the body of this big host fish worth the risk of being eaten for dinner? For the host fish, deciding to permit a pesky little fish or shrimp intimate access to its body has to be similarly difficult. Piscivorous hosts naturally feed on pesky, little bite-sized fishes, and for these the decision is even more difficult: Do I have a snack, or do I allow some small fish to wander about nipping at my fins or, worse, invading my gill chambers? Allowing cleaners access is logical only if the host appreciates that the cleaner will clean rather than take chunks of fin, gill, or other tissue as its own food. Claiming that this is all instinctual behavior selected through thousands of years of evolution does not solve this dilemma. Both cleaner and host face risks every time they engage in a cleaning experience.

Let me clarify that last point. Dismissing a challenging observation of behavior by saying that it is controlled by instinct does not explain its origin or its repeated performance by different individuals. For cleaning to have evolved as an instinctive behavior, cleaners and hosts had to come together despite the evident risks, and to do so many times; for cleaning to occur today or tomorrow, individual cleaners and hosts have to again come together. The risk may have been reduced by the ritualization provided by instinctive control, but it is still there.

Cleaning interactions are most elaborated, and the cleaning stations are most conspicuous, when the cleaner is *Labroides* (the handful of *Labroides* species are all about 10 centimeters in length and patterned in quite similar versions of horizontal blue and black stripes). Not surprisingly, scientists interested in cleaning interactions on coral reefs have mostly investigated cleaning by *Labroides* species, although there are also studies of Caribbean cleaning stations (staffed by gobies) and a few observations of cleaning shrimps.[18]

The conventional picture of the cleaning station—a permanent location on a reef—used by two or more *Labroides dimidiatus* and visited by client fishes belonging to a wide array of species, who line up patiently waiting to be cleaned like so many people waiting at a barbershop, has been reported so many times that it is likely embedded in the consciousness of anyone interested in coral reefs. Yet think about what that picture conveys. Small blue and black striped wrasses spend much of their day at a particular location, where they service a continuous succession of client fishes, swimming up to each, moving about close to or touching its body surface, nipping at minute ectoparasites on its skin, in the folds or at the bases of its fins, in its mouth and gill cavities. Client fishes come to the cleaning station, patiently wait their turn, then pose for the cleaner, raising their fins, resting immobile in the water column, flaring their gill cover or opening their mouth as necessary to allow the cleaner passage. Then, after being cleaned, the client swims away, his or her place to be promptly taken by the next in line. How did such a conspicuous pattern of interspecific interactions ever evolve? What is its significance to the fishes? And how do the participants manage this intricate dance?

Early research on cleaning symbiosis focused on describing the scene, recording the number of fish and of species cleaning or being cleaned, logging the food items in cleaner stomachs, and noting any apparent signals used to coordinate behavior of cleaner and host. The impact of cleaning on the fish community was explored in experiments in which cleaners were removed from reefs to see if any changes occurred. For Limbaugh, there was no doubt that cleaning symbiosis was a wonderful example of interspecific reciprocal altruism, in which each partner acts to improve the life of the other—a perfect mutualism in which the removal of parasites benefits the host while feeding the cleaner. That very neat story has become a bit more complicated over time.

Experiments to demonstrate the importance of cleaning to reef fish communities have yielded mixed results. Limbaugh removed all cleaners from two small reefs in the Bahamas. He observed a rapid loss of other fish species over a few days and a deterioration of condition of fish that remained (they developed fuzzy white blotches, swelling, ulcerated sores, and frayed fins, all of which Limbaugh attributed to effects of ectoparasites). But two larger, lengthier, better-controlled experiments in Hawai'i resulted in no observable changes in community composition or abundance of ectoparasites. A subsequent, even longer experiment on the Great Barrier Reef, which kept reefs free of cleaners (*Labroides*) for eighteen months, revealed losses in abundance and diversity of more mobile species but no effects on more sedentary species such as territorial damselfishes. More recent reports based on studies on small reefs kept free of *Labroides* for more than a decade have revealed reductions in abundance and diversity of host fish species, reductions in growth rate of fish that remain, and, most recently, changes to the gnathiid parasite community that resulted in greater rates of infestation than occurred on reefs with *Labroides* present.[19] It is clear now that there can be effects of cleaner removal on the gnathiid parasites (small isopod crustaceans), though perhaps not on others, that ectoparasites reduce condition of host species, and that fish species vary in their susceptibility to the parasites. Yet, removal of cleaners does not automatically result in a loss of host fish species from a reef. In other words, the removal of ectoparasites is helpful to a fish but apparently not an urgently important task that cleaners fulfill—all of which just makes the fact of cleaning symbiosis that much more amazing.

Dietary studies of *Labroides* have also revealed that Limbaugh's reference to feeding on "parasites and necrotic tissue" overstated the degree to which cleaners clean. *Labroides* feeds on gnathiid isopods and flukes but also on mucus, flesh, and bits of fin. They appear to prefer the mucus and flesh! In places and times where parasite loads are relatively low, *Labroides* continues to service hosts by providing tactile stimulation—most fish apparently like to be gently tickled—but act essentially as ectoparasites themselves while doing so. The altruism is a wee bit strained. Yet still cleaning symbiosis continues to occur.

The most recent research on cleaning symbiosis includes a number of studies that look closely at the behavioral interactions between cleaners and hosts. Most of this work has been done on *Labroides*, but Isabelle Côté of Simon Fraser University, Canada, and colleagues have done such studies on Caribbean cleaning stations operated by gobies of the genus *Elacatinus*. Much of the recent *Labroides* work has been done in Australia by a team including Redouan Bshary and Alexandra Grutter, University of Queensland. Grutter, uniquely, approaches cleaning symbiosis from the perspective of the ectoparasites, especially the gnathiid isopods. In these recent studies, cleaners are characterized as veritable Machiavellis who tantalize client fishes with gentle tactile stimulation, and some removal of parasites, in order to get close enough to gain the odd mouthful of mucus, scales, or flesh along with the parasites.[20] In other words, cleaners regularly cheat. They only offer what looks like altruism.

If cleaners regularly cheat, the whole question of how cleaning symbiosis developed and persists becomes even more mysterious than ever. If cheating happens too often, clients will cease visiting cleaning stations and may even start to cheat themselves, consuming cleaners in the process. (While cheating by cleaners is quite frequent, eating of cleaners by clients is rarely seen.)

When a *Labroides* cheats and takes a bite out of its client, the client responds with a sudden involuntary twitch called a jolt. This makes it possible to observe frequency of cheating during a cleaning episode, and studies have shown that *Labroides* cheat at rates that are client-dependent. Larger clients and clients of species that tend to have heavier ectoparasite infestations are favored because they offer better feeding opportunities; predatory clients are clients that might harm or eat cheating cleaners. *Labroides* cleaning these larger, well-loaded, or piscivorous clients tend to cheat less frequently than when they clean other species. Even more surprisingly, *Labroides* tend to cheat less often when other potential clients are present watching the cleaning performance! There is also evidence that individual *Labroides* recognize at least some individual clients and prefer to clean them rather than other fish of the same species.[21]

Client fishes are not passive partners in a cleaning interaction, but there has so far been less study of what their behavior means. On entering a cleaning station, a client will frequently pose, erecting its fins and resting immobile in the water column, usually in a head-down posture. This happens a bit over 50 percent of the time for clients entering a Caribbean *Elacatinus* cleaning station. It has long been supposed that posing is an invitation to clean, but Isabelle Côté found the story not quite that simple.[22] Individuals that posed were more likely to be cleaned, but there was lots of variation in the cleaners' responsiveness, as well as among client species in the likelihood of posing. Species that were more likely to pose were not necessarily more likely to be cleaned in Côté's study, although George Losey had shown that individual clients that had been deprived of access to cleaners were more likely to pose to *Labroides* than individuals that had recently been cleaned. Posing does seem to be an invitation, but it is an invitation that may or may not be responded to by the cleaner.

Putting all the work on cleaning symbiosis together, we see a dramatic interspecific drama, involving multiple individuals of many species interacting in a way that permits extreme closeness of small, tasty cleaners and larger, often predatory clients. There are food rewards for the cleaners, and comfort rewards (all that tickling) plus possibly a reward in terms of relief from ectoparasite infections for the clients. But the interactions are a complex set of behaviorally mediated negotiations by animals that may at least sometimes know each other as individuals. Given the number of cleaning encounters per day, and the large number of species involved, there is clear evidence, at least for cleaners like *Labroides*, of considerable ability to discriminate individuals and make judgments concerning alternate clients.

It's fair to say there remains a lot we do not know about how the residents in reef neighborhoods interact. What we do know reveals these neighborhoods as socially vibrant communities filled with negotiation, collaboration, and some lying and cheating, all conducted, not just across species boundaries, but among species of radically different type to build a complex society. That cantina on Tatooine is a very pale reflection of a cleaning station, and cleaning stations are just one particularly memorable part of what happens on coral reefs.

SIX

Wonderful Surprises

How Do Coral Reef Communities Assemble?

Shakespeare wrote, "There are more things in heaven and earth, Horatio, than are dreamt of in your philosophy." When ecologists dream, they are a lot like Horatio. Too conservative . . . and the natural world is forever granting surprises. Those surprises create the intellectual excitement that makes the life of a research scientist so rewarding.

One of the biggest surprises in my life has been to learn how reef fish communities are assembled. By "assembled" I am referring to the rules that govern the combinations of species and the numbers of organisms living at a place, as well as the changes in these through time. Perhaps I should say "presumed rules," but I think it is obvious that ecological communities of reef fishes are structured in particular ways and that we do not see various hodge-podges of creatures happily milling about at different places on a reef. Since there appears to be structure, there should be rules, and ecologists should try to sort out what these rules are.

Like most biologists who grew up at the time I did, I picked up a series of ecological rules in the course of my education. These were seldom precisely specified rules that could be reduced to neat equations—they were more like something you might read in an old *Farmer's Almanac*—but those were the rules we had. We justified the lack of specificity as being a consequence of the fact that nature is complicated.

The rules were declarative statements: "An ecosystem will contain a smaller biomass of carnivores than of herbivores than of plants," or "Species coexisting in an ecological community will differ in one or more aspects of their niche," or even "Tropical systems will be richer in species than temperate systems." In many cases the rule described a pattern flexible enough that any of several quite different processes might be responsible for its existence (the one about tropical richness has been "explained" by at least eight different hypotheses, most of which are compatible with each other, making any attempt to sort them out rather difficult).

As a student I also assimilated the widely prevailing assumption of the time, that ecological systems are usually at some definable equilibrium state rather than in the process of changing from one state to another. Many people know this rule as the balance of nature. This assumption of stationarity meant that ecologists could largely forget about time and change, because the forest is the forest is the forest, or alternatively, a reef is a reef is a reef. We all knew that ecological systems had changed greatly over geological time, but we did not think it important to worry too much about directional change in structure or organization over the shorter time scales of interest to us—decades to centuries.[1]

The rules governing the assembly of communities have not noticeably tightened up in the years since I began my career, although our attitude to stationarity has changed a lot. Ecologists still prefer verbal statements to equations, much to the consternation of physical or chemical oceanographers who find themselves collaborating with us. And, yes, we continue to expect that the ecological world is governed by a set of straightforward rules, if only we could figure them out! I've come to believe that we remain unable to quickly specify the rules operating in a particular community because we continue to expect simple rules. We know that the balance of nature is a myth, yet we still subconsciously expect ecological systems to be unchanging and to be governed by simple rules. The world around us does not operate by simple, global rules, and that is certainly true for coral reefs and other complex ecological systems. I did not know that when I was a student.

When I arrived in Honolulu in 1964, my view of the world had already been strongly shaped by conventional niche theory—to survive together, coexisting species would have to differ in their ecological requirements. Otherwise, superior competitors would outcompete inferior ones and efficient predators would wipe out less capable prey. Only groups of species that differed in ways that gave each a good chance to survive and reproduce would occur together over any length of time; these would be the groups of species we would find living together in any particular place. I knew that not all ecologists believed this rule was correct, but I also knew that the great majority did. I also recognized that this rule had contributed much to ecological thinking outside academia, especially when it came to managing wildlife or national parks. I certainly was not about to challenge a central tenet of my field of science. In fact, I intended to provide more evidence of the generality of this rule, by carefully examining some co-occurring species of reef fish to find out the ways in which their niches differed. In this way I would add a couple of bricks to a growing edifice of ecological theory.

Looking back, I am embarrassed by the lack of originality in my thinking—I was content to collect evidence to support, rather than devise experiments to challenge, the prevailing hypothesis (that niche differences explained why certain groups of species reliably occurred together). That is not the way to build scientific understanding.

And so, one morning, I wandered into Bill Gosline's office, and he quietly told me about moray eels, as I recounted in chapter 4. Gosline was not an ecologist; as an ichthyologist, he appreciated the great variety of morphology of fishes, how that morphology predisposed each species to behave in particular ways, and the core of evolutionary theory that explained patterns of diversification to produce a rich fish fauna of many different yet related morphologies, each well adapted to its particular way of life. He may have thought that the variation among morphologies positioned each species within a different niche, but he also recognized that there were many examples of closely related species which had morphology that was very similar—different enough to make them recognizable as different species, but not really very different at all. He thought my naive idea of measuring niche features or ecological preferences among superficially similar species was not likely to result in anything particularly noteworthy. And a Ph.D. thesis that carefully reports only the obvious does not make for a promising science career. Thankfully, I turned my attention to habitat preferences and the behavioral mechanisms that might cause juvenile reef fishes to choose the correct habitat in which to live.

But the question of coexistence of species stayed with me. How were rich tropical ecosystems structured? What were the assembly rules? Many years later, I still don't know what those rules are, but I do know that they are many times more complex than the simple rules I or anyone else was thinking about in the 1960s.

♦♦♦

Once in Australia, casting about for new questions to ask of coral reefs, I turned my attention to the territorial damselfishes as described in chapter 5. Here were animals that pugnaciously defended their small territories from an army of potential invaders, including neighboring damselfishes (fig. 17). They got their algal food from within the territory, acting in many cases in ways that some ecologists had interpreted as farming their algal food. And yet I was seeing individuals of different species occurring side by side, defending their small territories from each other. Watching their real estate transactions over time and undertaking experimental removals to see whether the particular patch of space would be taken over by the same species or by one of the others, I was asking the fishes to tell me whether they used the same type of space or subtly different kinds of space. If two species used the same type of space, they were identical with respect to that important niche dimension (important because they were fighting over it). If they used subtly different types of space, the niches differed.

Initially, I expected that I would find subtle differences among species, in accordance with accepted theory. As time went on, my doubts grew. Many of the eight species I studied certainly had clear ecological differences. Some were strictly limited to shallow reef flat sites, others to the reef slope. But some species were occurring side by side in the same rubble patches and doing so successfully over years. Furthermore, I saw newly settled juveniles occupying tiny sites on the borders of territories of animals much larger than they were. These juveniles used the topography to their advantage, ducking into holes too small for the larger fish to enter rather than fleeing when attacked. Sometimes these juveniles grew up, gradually capturing more

Figure 17. Numerous species of damselfish are territorial herbivores, and many will aggressively defend their homes. *Stegastes nigricans* is a widely distributed Indo-West Pacific species and a particularly belligerent defender of its territory, usually within a thicket of *Acropora* sp., where it maintains a lush garden of algae. This Great Barrier Reef individual, perhaps 12 centimeters in length, has no hesitation attacking a diver fifteen times its size. Photo © Luiz Rocha, California Academy of Sciences.

and more territory from their neighbors as they did so. I began to realize that owning a site puts a fish at an advantage in any fight over access to that site, that a small owner can repel a large intruder.

At some point in this study I also realized the importance of the life cycle of reef fishes to this struggle for space among damselfishes. The success of an individual damselfish in capturing and holding a territory makes little difference to the success of his or her offspring, although good success likely means that he or she will produce more of them. Every generation, the larval fishes must survive a month or so at sea, return to a reef, and find a suitable place to live. Furthermore, the vast majority of the millions of larvae that hatch will die before the larval life ends. If suitable space is in short supply, the first fish to arrive at an available reef site, rather than the particular species best suited to it, is the one that will capture it and hold it perhaps through a long life. Species that are approximately equivalent in their ecological requirements can coexist indefinitely.

Put another way, the theory underpinning our rule that coexisting species must have different niches assumes that species compete in a closed environment in which successful adults produce larger numbers of offspring that in turn are successful because they inherit successful traits. Gradually, the best-adapted individuals come to dominate, and species must have ecologically important differences in order to occur in the same environment. But coral reefs are decidedly not closed. Every newly hatched organism is out in the open ocean, must survive the rigors of that life, and must then find its way to a reef and to the correct habitat there.[2] In essence, the competition for space between neighboring damselfishes has zero effect on the chance of success of larvae in the next generation. I saw that, under these circumstances, damselfishes could be competing strongly for a limited supply of suitable space on the reef but that winning more than 50 percent of the time (being a superior competitor) would have a negligible, if any, effect on the likelihood of future offspring successfully capturing space.

I further convinced myself that this openness was important by building a simple computer model in which three species of fish competed for a set of living sites but exported their offspring. Sites became available as resident individuals died. Recruitment to those sites occurred from an assumed global pool of larvae, with the proviso that as a species became more common as adults occupying living sites, the proportion of larvae in the global pool increased, but in a nonlinear fashion, as would be expected if larval mortality was very high and the set of living sites being examined was small relative to the universe of such sites that provided larvae to the pool. That is essentially the set of conditions I was observing when I watched small groups of fish fighting for territories in small patches of suitable habitat.

The results of my modeling surprised me: ecologically identical species persist for long periods in this system (I routinely ran the model for two hundred to five hundred generations).[3] More surprisingly, combinations that pitted two identical species against a competitively inferior one persisted for ecologically long periods of time unless the ecological differences in fecundity, longevity, or both were quite large. These results convinced me that coral reef species that were moderately similar ecologically could persist together for long periods of time, even though their individuals competed strongly for such resources as living space.

Looking back at the work I did in the mid- to late 1970s and remembering how I formulated ideas and undertook research to test those ideas, I am fascinated at how my thinking changed. From the student who expected the world to be organized according to niche theory, I became a scientist who doubted that niche theory explained very much about anything. And yet I continued to believe that ecologically similar reef fishes were competing for scarce resources on the reef (codified as space resources in the case of the damselfishes). I was still wedded to the belief in a world at stasis with ecological systems at or very near an equilibrium state with each component species limited by its competitors, or perhaps predators, and all resources being fully used. My departure from the conventional ecological understanding at the time was merely to suggest that very similar species could coexist for many generations because the world was an open environment.[4]

In those days, a time before Skype or email or even intercontinental phone calls casually made, I was making trips to North America every year or so, partly to visit family and partly to keep myself informed on what was happening in my science in that other part of the world. On each visit I would attend an ecological conference and likely also visit one or more universities to present seminars on my recent work and learn from colleagues about the work they were doing. I found my ideas about coexistence were stirring up controversy among ecologists. My departure from conventional views was controversial, and the majority of my colleagues rejected it. Either Australian coral reefs were peculiar or Sale had somehow got it wrong. I remember one particular occasion quite vividly, although I cannot place the precise mid-1970s year.

I was invited to present a seminar at the University of Michigan. On the appointed day I faced a large crowd in a well-filled lecture hall, and I told the story of my damselfishes and my interpretation of how they were successfully sharing space. When it came time for questions, there were lots of them, and I did my best to fill in the details that people asked about. Then one individual, seated almost in the center of the audience, rose to his feet. He was about my age, likely a young professor, and his rising to his feet, instead of simply asking his question while seated, was definitely intended as a threat. He was going to ask something difficult, and score points in the process. I steeled myself.

His first words, to the best of my memory, were, "My name is Steve Hubbell, and what you have just told us is impossible." A rustling went through the audience. "No, it isn't," I said. "Why do you say so?" "Because when two species compete, the superior competitor must inevitably outcompete the inferior." "Yes," I said, "that is correct in a closed environment, but we are not dealing with a closed environment here."

We went on for several more exchanges, and finally I suggested, "Maybe we can talk later?" He readily agreed, and we arranged to meet in his office the following day. That follow-up conversation was long and wide-ranging, and I now recall few details. But I do remember that I could not convince him that what I was suggesting was even a possibility. Nor did he convince me that I was wrong. We parted on friendly terms but agreeing to disagree. Several years passed, and I was surprised one day to see a new paper by him in *Science*, concerning rain forest trees in Panama. The forests he worked in were exceptionally rich in species (as is typical for many rain forests); each individual tree of a species had about fourteen different species among its twenty nearest neighbors, and any two trees of one species had only about

30 percent of nearest-neighbor species in common. In other words, each individual of a species was confronted with a different mix of nearest neighbors, and in trees it's the near neighbors that compete most intensely. They all compete for space in the canopy. In his paper he presented a model of the coexistence of these many tree species that relied on random opening up of the canopy (through treefall) and random germination and growth of one or more seedlings to fill that gap.[5] The result was that the many species of tree shared the available living space in a way very reminiscent of my damselfishes, and in a decidedly nonequilibrial but persistent way. I chuckled when I saw it. Years later, when I met Steve again, we laughed about our first encounter and about how far his ideas had shifted.[6]

♦♦♦

During the time I was watching damselfishes defend their territories, I began a quite different study to further explore how reef communities were assembled. This began with a two-year-long study at Heron Island and was followed by a ten-year-long project at One Tree Reef.[7]

Even quite small patches of coral will be occupied by little fishes that are resident over months to years, particularly when that coral patch is separated from other reef by a few meters of open sand. In the pilot study at Heron Island, I used single living or dead coral heads fastened to concrete blocks, but at One Tree Reef I used substantially larger, though still small, coral patches as a proxy for much larger coral reefs. I used these rather small pieces of reef because they contained groups of fish that were large enough to contain many species and yet small enough that it was possible to make reasonably accurate counts of all daytime active species without needing to capture or immobilize the fishes.

In the pilot study, my technician, Rand Dybdahl, and I set out ten living and ten dead coral heads in a sandy patch on the Heron Reef flat. We then proceeded to remove all fish from each of sixteen coral heads (eight living and eight dead) every three or four months, for six collections over two years; two dead and two living coral heads were left uncollected throughout the experiment. I presumed that, although I had chosen coral heads of relatively similar size and shape, each was subtly different and might provide different kinds of space and other resources for resident fishes. Certainly, the dead coral heads were quite different to the living ones. I expected that if the various fish species had subtle but important differences in their requirements, I'd find (1) that dead coral heads would collect a different set of species than would live coral heads, and (2) that individual live coral heads, or dead coral heads, would tend to collect slightly different mixes of species from other heads of the same type.

In the larger experiment at One Tree Reef, we used twenty small coral patches and decided to not disturb the fish in any way, except by looking at them every three to four months to get as accurate a count as possible of which species were present and how many fish of each there were.[8] My hypothesis was that, since every small coral patch was different in many ways to the others, if reef fish were habitat specialists, the patches would be occupied by different mixes of species of fish, and these differences would persist through time, even as individual fishes died and were replaced by new recruits. If, however, fishes did not discriminate among coral patches, such a pattern of distinct faunas would not emerge. By the time I commenced this experiment, I would have been very surprised if I found fine discrimination among habitats—local-scale patterns of distribution of reef fish were, in my mind, far less precise than conventional niche theory might predict.

The results lived up to my expectations in several ways. In the pilot experiment, we collected 630 individuals belonging to 56 species of fish from the twenty small coral heads. On average, living coral heads supported more fish than did dead ones, and 5 common species clearly preferred live coral. But overall, there was little pattern to which species occurred in which coral head.

This lack of pattern was even clearer in the ten-year study of natural coral patches at One Tree Reef. These larger units of habitat supported larger numbers of fish, and over the ten years we recorded 50,124 individual sightings of 141 species of fish, with individual coral patches supporting 125 individuals belonging to 25 species on average at any given census. With ten censuses of each patch over ten years, it was possible to explore how assemblages of fish varied among the coral patches in several different ways. We found that every coral patch was home to many more species over the experiment than it held on any single census. We found species that varied dramatically in number from census to census and species that were consistently present in similar numbers each census. We found that the number of individuals present on single coral patches was related to the size of the patch, but to little else. In other words, fine details of the habitats a particular coral patch provided, even such seemingly important details as the amount of living coral present, did not help us predict the number of individuals, the number of species, or the particular species present either at a single census or over the course of the experiment. And we could not identify subsets of species that tended to occur together or that tended to avoid each other in distributing themselves across the coral patches.

In my opinion, the results pointed to a rather simple rule concerning the assemblage of reef fishes at lagoonal coral patches. Of all the fishes that occur at One Tree Reef, several hundred species occur within the lagoon, and of these at least 141 species will occupy small coral patches. Fish of these species complete larval life and settle to reefs, finding suitable patches of reef habitat across the lagoon. They do not discriminate finely among coral patches, and every small patch is suitable habitat for many more species than it could contain at any one time. What it holds on any one occasion is little more than a random sample of the species potentially available. There are unlikely to be any complex assembly rules for reef fishes that could explain the details of their patterns of distribution across patches of available habitat. There are no tight structural commandments that determined which species occurs where—life for reef fishes really seems, in this respect, pretty much a lottery.

♦♦♦

How did these ideas of lottery competition fare as time and research by others went on?[9] I know that my work helped raise doubts concerning the realism of niche theory. There was grudging admission that, at least in species-rich habitats like coral reefs, or in places where the young individuals lived in a different environment, with its own ecological demands, something like lottery competition might occur. And lottery-type interactions do seem important in a number of plant communities where competition for space can be very important. But my confidence, in the mid-1970s, that I had probably solved the coral reef question (how so many species coexist) gradually faded. I still think lotteries are more important than many other ecologists do, but I now see them as kicking in only occasionally, when resources become scarce.[10] I believe this because of something else we discovered back in the 1970s that was even less expected than the idea that similar competitors could coexist.

♦♦♦

Of course, all the time I was busily tracking the fishes that occupied my coral patches, my graduate students and I were starting to look into the process of settlement of larval fish and their recruitment into local populations on the reef. And we discovered something really unexpected.

The prevailing view in the 1970s, as I said earlier, was that ecological systems existed close to some state of balance, with the number of species, the particular species present, and the numbers of individuals of each pretty well set by the competitive and predatory interactions among the species concerned. The prevailing view of coral reef systems was that the individual fishes present competed strongly for shelter space on the reef. Space was the resource in shortest supply and the one that drove community composition. One surprise that the study of coral patches revealed was that when you censused carefully, counting all the fish present, the number of individual fishes living on a coral patch would fluctuate from census to census. And it did not fluctuate in the same way across all coral patches—some months numbers were high on some patches, other months different patches held large numbers. Indeed, over ten years, numbers on single coral patches varied at least twofold, and on twelve patches numbers varied tenfold. If space was a resource in short supply, how could this happen? Clearly, on some occasions space was a lot more in demand than at others![11]

Peter Doherty joined my lab for a Ph.D. following completion of his M.Sc. degree in New Zealand. There are many tales I could tell about Peter, who went on to do reef fish research of major importance during his long tenure at the Australian Institute of Marine Science, but I'll be kind and tell just one. It is pertinent to this issue of whether reef fish populations are limited by availability of living space on the reef. Peter chose to look further at *Pomacentrus wardi*, one of the territorial damselfishes I had studied, hundreds of which occurred on the large patch reefs scattered within the One

Tree lagoon.[12] Peter was a good deal more energetic than I ever was. Where I had been content to do removal experiments extracting three or four territory holders from within a patch of habitat at Heron Reef, thereby liberating two or three square meters of habitat, Peter embarked on a bit of a blitzkrieg, removing all *P. wardi* from each of four large patch reefs in his first experiment: 29, 31, 51, and 69 fish speared and removed.[13] He then waited and watched for the rapid recruitment of settling larvae that we all knew was certain to occur. Two years later, his patch reefs contained 12, 15, 14, and 28 fish, respectively, an average of 39 percent of the number originally present (and because these were all young, small fish, just 26 percent of the starting biomass). The rates of settlement to these empty patch reefs were no higher than to undisturbed, occupied reefs nearby. What was going on? Peter became a little embarrassed about what he had wrought. Why hadn't these so-called empty patch reefs filled rapidly with newly settled recruits? Had he done something to permanently change the reef?

Over time, the patch reefs did fill back up again, but we had learned that the real world of coral reefs was even more different from what we had all been assuming. Space on the reef was not always in short supply, even if damselfishes fought over pieces of it. And with this, the interest in settlement and recruitment of fishes on coral reefs took off in Australia and on reefs around the world.[14]

SEVEN

How Nemo Found His Way Home

How do reef fishes know where on a reef they should be living? Such a simple question, yet one that has occupied me off and on throughout my career. I still do not know the answer. With a handful of exceptions, they leave the reef environment for the open ocean at spawning or immediately on hatching and spend typically a month at sea before returning to a reef environment. With no or little prior experience of the reef, how on earth do reef fishes know the type of place on a reef they should choose to live in? This same question can be, and is being addressed to corals, to crustaceans and mollusks, indeed to most types of reef creature because nearly all of them also have a pelagic larval phase. Still, this chapter will be another one about fishes because we have been able to advance much further toward answers with the fishes than with any of the invertebrate groups. Also, I know a bit more about the fish story—if you have a fondness for cowries or giant clams, or for spiny lobsters or branching acroporid corals, use what I tell you about fish to start thinking about what might be possible with these creatures, because they all have pelagic larvae, too.

Although we have yet to answer the big "how do they know" question, we have learned a lot about how reef fishes get back to the right places at the end of larval life. In truth, the big questions are usually ones that scientists reserve for pondering over beers, rather than the ones they use in writing grant applications—often they are fundamentally impossible to answer using science, and this question is one of the almost impossible ones. But the fish do know, and many reef scientists have struggled long and hard to find out how they know.

When I was a graduate student, we knew relatively little about the larval lives of fishes, particularly if you consider the behavioral aspects of that life. We usually saw larval fish only after we had extracted them from the ocean by towing a large-diameter, fine-meshed net behind a boat or research vessel—fish trapped in such a net get pushed down to the rear end (the cod end), where they get more or less mashed together with all the other plankton and from where they are dumped unceremoniously into a sorting tray or directly into formalin or alcohol. Any behavior they might still possess ends at that point, and even the fine details of their morphology tend to get lost as fragile fins and filaments get broken. In any event, the limited research done on behavior of larval stages of temperate fish such as cod had shown them to be feeble creatures not even capable, for the most part, of swimming against the usual currents one finds at sea. Why should reef species be different?

As a beginning Ph.D. student at the University of Hawai'i, I was drawn to the early life of the manini (*Acanthurus triostegus*), a pale tan surgeonfish (family Acanthuridae) with darker vertical bars, very common on Hawaiian reefs. The great tropical fish biologist J. E. "Jack" Randall had completed his own Ph.D. about ten years earlier in Bill Gosline's lab, and 50 percent of his dissertation, which I read avidly, cover to cover, concerned the biology, including reproduction and early life history, of the manini. Jack had established their spawning season (about nine months long), that they spawned monthly during that season, and that their offspring apparently were pelagic larvae for about two and a half months before recruiting to the shallowest parts of coral reefs and inshore tide pools, where they spent their juvenile lives. I say "apparently" because Jack deduced the length of the larval life from the mismatch between the timing of spawning (as judged by the state of ovaries of fish collected in different months) and the timing of recruitment to the reef as judged by when he was able to find newly arrived manini in the shallows. These manini turn up at night around spring tides in each of about nine months of the year. They are beautiful, transparent creatures about 2.5 centimeters long with mouths still adapted for feeding on midwater plankton. Jack showed by simple aquarium observations that newly arrived, transparent fish began a substantial metamorphosis immediately on arrival in the shallows, developing pigments and scales while tripling the length of the intestine and modifying mouths and teeth to facilitate an algal browsing lifestyle. The metamorphosis is complete in about three days, but by their first morning in shallow water, manini have begun to develop scales and pigments. Jack also showed that larval manini, collected in open water by shining a light and dip netting them from beneath it, would undergo transformation if kept in a bucket but would remain transparent if kept in an all-glass aquarium suspended so that there were no solid-appearing (opaque) surfaces accessible to them. I can confirm that ready-to-settle larval manini are large, conspicuous, beautiful fishes, seen at night over shallow reef flats and easily distinguished from juveniles that have been on the reef for twelve hours or more. I have never seen a larval manini above a reef during the day or collected one at night that was not in the process of swimming toward shallow water and commencing its metamorphosis into a juvenile.

Think about it for a moment: manini eggs are released and fertilized in midwater, high above an outer reef edge, usually on an outgoing spring tide. They hatch within hours and then have an open ocean experience lasting ten weeks. For at least half of that time they are so small that they cannot possibly swim against a current. Somehow, at the end of that time, they appear at night on coral reefs swimming rapidly toward the shallowest, most inshore portions, where they will spend the next year of life, slowly moving out toward deeper reef waters as they grow larger. Growth is initially rapid in surgeonfishes, so they reach adult size within three to five years; maximum longevity can be thirty to forty-five years.[1] How on earth does the larval manini know that it should find a reef and move into the shallowest part for its juvenile life? Given that it does know, how does it make this journey, or is it only the lucky few that stumble back into reef waters? Suggesting that the knowledge is instinctive is not a real answer to the first question, and even if knowledge of what constitutes the correct habitat at each life stage is under strong genetic control (which is what calling it instinctive means), that hardly explains how the journey is accomplished. To begin with, how does the larval fish, far out in the open ocean, even know in which direction a reef lies?

Fascinating questions. Exceedingly difficult to answer. In my own doctoral studies, I focused on the juvenile life stage and asked simpler questions: Did young juveniles prefer shallow to deep water? Were there other attributes of the habitat, such as suitable shelter, that influenced their habitat choices? These questions were difficult enough to tackle, within the capacity of a young Canadian struggling to understand something about coral reefs. Science really does grow by baby steps.

The questions surrounding the larval lives of manini also surround the larval lives of virtually every other coral reef fish. Except for about a half dozen species around the world, every coral reef fish has a pelagic larval stage (fig. 18). About half of them, like the manini, have eggs spawned in midwater and usually on outgoing tides; the other half deposit their eggs into nests or carry them in the mouth or in a pouch possessed by one parent, but even these, on hatching, immediately rise up into the water column and are taken out to sea on the tide. Larval reef fishes are simply not found in coral reef environments, except for that exceptional half dozen species that rear their eggs and care for their larvae as do many freshwater fishes, avoiding the pelagic phase entirely.

♦♦♦

The story of the larval life of coral reef fishes, and particularly the story of how they return to reef habitat, has intrigued many reef scientists throughout my career, and I've been able to watch as we collectively applied new, ever more sophisticated techniques in figuring it out step by step. It has been one of the fascinating stories around which my own career kept getting entangled, an amazing story that reveals another way in which coral reefs are incredible systems. I still have difficulty believing some of the facts we have discovered.

In the 1960s, we assumed that larval fishes would be largely incapable of swimming against oceanic currents; instead, their larval journeys would be ones determined by the currents themselves. One popular hypothesis then for how manini could get back to reefs around the Hawaiian Islands centered

Figure 18. Larval reef fishes can be spectacular photographic subjects, especially when images can be shot in the field at night using blackwater lighting techniques. Surgeonfishes, such as this blue tang, *Acanthurus coeruleus*, photographed in July 2016 in the Gulf Stream off Boca Raton, Florida, are particularly impressive because of the relatively large sizes they reach toward the end of larval life. Shimmering, transluscent skeletal elements are almost the only opaque parts, and yet within twelve hours of settling to a reef, the fishes become opaque and begin to take on their characteristic colors. Photo © Suzan Meldonian, https://www.niteflightphoto.com/.

on the frequent occurrence of large gyres southwest of each island and the possibility that larvae, entrapped within the circular flow, would be transported back to the islands at about the time they'd be ready to metamorphose into juveniles. After all, passive drift was the rule for the types of larval fish from temperate waters whose swimming behavior had been examined.

Water is a remarkably viscous liquid if you are very small (as all larvae are, at least at the beginning of their lives). Scientists interested in the biology of planktonic organisms have explored this feature in some detail and can talk endlessly about viscosity, swimming efficiency, and Reynolds numbers (R_e).[2] To grasp the larval fish's problem, we must remember that the Reynolds number varies with size of the fish and swimming speed as well as with viscosity of the water. A Reynolds number greater than about 300 is needed for the fish to be able to move easily through the water.

When R_e is less than about 300, the larva is trapped like a fly in honey; with an R_e more than 1,000, the larva swims freely through the water. A Reynolds number between 300 and 1,000 is a region of intermediate conditions—the water seems a bit oily. Now, for tiny larval fishes, 5 millimeters in total length, the universe is a vast, viscous fluid unless they can swim at about ten body lengths per second. For smaller larvae still, or for slower larvae, the universe remains permanently viscous, and swimming at all is energetically costly. All larval fish start life smaller than 5 millimeters long, and some never become capable of swimming ten body lengths per second.

To put this properly into perspective, on August 1, 2009, 193-centimeter-tall Michael Phelps set the world record for the 100-meter butterfly by swimming at 1.04 body lengths per second for 49.82 seconds—an amazing achievement. But were he to have managed 10 body lengths per second through water, it would have been sensational! For larval fishes a centimeter long it is possible to break through the $R_e = 300$ barrier at a speed of about 3 body lengths per second, and for larval fishes 2 centimeters in length the opportunity to swim freely comes at a Michael Phelps speed of 1 body length per second. Late-stage larval manini moving into shallow water are 2.5 to 3.0 centimeters in total length, and capable swimmers indeed.

Jeff Leis is a laconic American ichthyologist who has spent his career in Australia as part of an amazing ichthyological team at the Australian Museum. Jeff obtained his Ph.D. at the University of Hawai'i in the early 1970s studying the distribution of larval fishes offshore from O'ahu. Using fine-meshed nets that could be towed at fixed depths, and somewhat to his surprise, Jeff found patterns in distribution that could not be explained simply as due to effects of currents moving larvae about. Larval fishes of different species were found at characteristic but different distances offshore and at varying depths. Clearly, the fish were determining, at least partly, where they occurred in the open ocean. Jeff devoted much of his career to study of the ecology of larval reef fishes. In a chapter for my book *The Ecology of Fishes on Coral Reefs*, he had the temerity to proclaim that coral reef fishes were really all pelagic species that happened to have a reef-associated adult phase.[3] He was pleading with his colleagues to think more deeply about the larval phase that we all tended to ignore—important biology was happening out there in midocean.

Among his many accomplishments, Jeff tackled the issue of larval behavior head-on. Being a practical soul, he figured that if you want to find out what an animal is doing, you should go out there and see for yourself. So Jeff developed blue-water diving techniques that would put him in the water, far from any reef, with reef fish larvae that he would then observe and follow. Jeff is about as tall, though not as fast a swimmer, as Michael Phelps, so I find it hard to believe that the larvae were unaware of his presence. Still, he observed strict protocols to prevent inadvertent herding, chasing, or startling, and he obtained the first direct measurements of typical swimming behavior by late-stage larval reef fishes. His work left no doubt: these larvae definitely could swim. And his measurements of swimming speed in the wild conformed nicely to measurements of maximum swimming speed obtained by placing larvae into flumes in the lab and forcing them to swim against a measured current (Jeff was involved in many of these lab experiments, too). The field measurements were typically about half as fast as the maximum speeds obtainable in lab flumes, but some observations included periods at speeds approaching the lab maxima. These fish could definitely swim faster than prevailing currents in the area, and nobody would expect fish in the wild to be swimming always at their maximum speed.

The work by Jeff Leis and others on swimming capabilities has revealed that virtually all species of reef fish tested swim far more efficiently than do the larvae of cod, plaice, herring, flounder, and relatives living in temperate waters. The reason appears to be simply because these are very different fishes. The bulk of coral reef fishes belong to the order Perciformes, the perchlike fishes, a large group of advanced, actively swimming species only quite distantly related to the Gadiformes (cods), Clupeiformes (herring), or Pleuronectiformes (plaice, flounders). Other important differences may be larger size (many coral reef species have larvae that become 1–3 centimeters in total length before settlement to a reef) and warmer water (viscosity is lower at warmer temperatures).

And how fast can reef fish larvae swim? That depends partly on the fish, because reef fish larvae come in many shapes and sizes. The challenge of learning about how fast reef fishes could swim had to await technical advances: the invention of effective light traps that would collect late-stage larvae in the open ocean, and advances in aquaculture that permitted rearing of many reef fish species through their larval and juvenile stages in aquaria.[4] These innovations were developed by bright, chiefly Australian, scientists seeking ways to get hold of larval reef fishes alive and in good condition—Peter Doherty played with ideas for light traps as a student in my lab (when he wasn't spearing damselfishes or turning dead *Tridacna* shells into curios) but got serious once he was a scientist at the Australian Institute for Marine Science with plenty of government funding and a full machine shop looking for work to do. His glossy Plexiglas light traps, about a meter high and looking vaguely like a lunar lander, became the gold standard for a time, as well as yielding interesting data on the timing of return of larval fishes to reefs.

By the late 1990s, Dave Bellwood from James Cook University had teamed up with Jeff Leis to explore swimming in larval fishes. Dave, who has become ever more mad-scientist-like in appearance as he has aged, has always bubbled over with crazy ideas. He pioneered the use on reef fish larvae of a technique for measuring swimming performance.

Most fish put into a current will automatically attempt to swim against it if any visual cues are available to tell them if they are moving backwards or forwards. They swim tenaciously, whether they are trout that live in streams or reef fishes used to poking about in a quiet lagoon. This habit is one of the things that seems to make a fish a fish. By putting single, late-stage larval fishes into a flume and forcing them to swim against a slowly increasing current, Dave measured what he called the critical swimming speed—that current speed at which the fish was just able to swim sustainably against it without giving up.

Working with Jeff, and with Ph.D. students Ilona Stobutzki and Rebecca

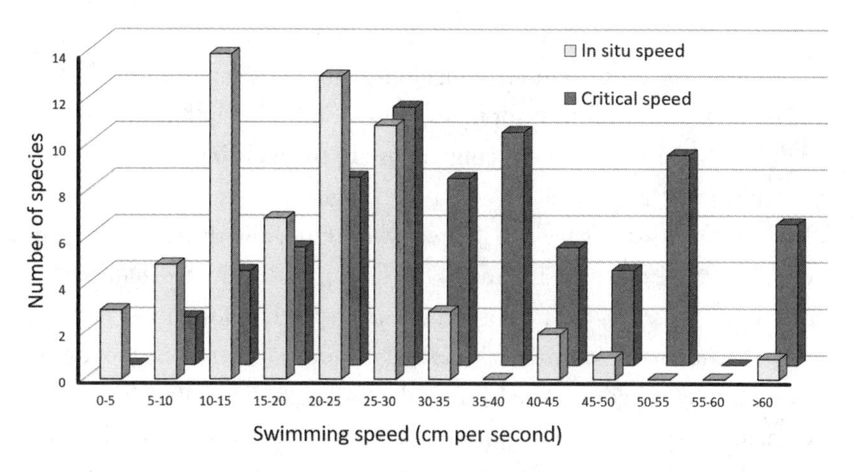

Figure 19. There is a broad range of swimming speeds among larvae of reef fish species with fish in the same family generally showing similar capabilities. This graph plots average in situ speed measured by following larvae swimming in the ocean (white bars) and average critical speed measured by swimming fishes at their maximum sustainable speed in a flume. Since most larvae are 1–3 centimeters in length, speeds shown here are mostly in excess of 10 body lengths per second. As the graph shows, most larvae perform at speeds equaling or exceeding average current speeds (about 13 centimeters per second) near Lizard Island where many of these data were obtained. Image based on data assembled by and courtesy of Jeffrey M. Leis.

Fisher, Dave obtained lab results for comparison with the field data Jeff was collecting. The results are surprising. Critical swimming speed varies consistently among species as we'd expect, but while some barely manage to sustain swimming at 5 to 10 centimeters per second, most reef fish larvae do far better than this, and the fastest among them achieve an astonishing 60 centimeters per second, or about 20 to 30 body lengths per second. To equal that, relatively speaking, Michael Phelps would have to complete his 100-meter butterfly in 2.5 seconds or less!

The manini and its fellow surgeonfishes turn out to be among the most competent swimmers, partly because by the time they are returning to reefs they have had a relatively long two to three months to grow and are relatively large (2–3 centimeters in length). The great majority of reef fishes can swim faster than prevailing currents in the locations from which they were taken, and the typical swimming speeds seen in the field are about half the maxima recorded in flumes (fig. 19).

A second unexpected result from the swimming speed studies was that larval reef fishes are particularly tenacious swimmers, capable of swimming at a good clip, continuously, for many hours, even without feeding. Some of the flume studies involved enticing the fish to swim against a current of just 13.5 centimeters per second (6 or 7 body lengths per second for a surgeonfish) until they finally gave up, stopped swimming, and drifted back against the screen at the downstream end of the flume. The time to exhaustion varied among species, as did maximum possible speed, but the most tenacious species swam many kilometers over many hours under these conditions. Among the less accomplished fishes, damselfishes, typically 1 to 1.5 centimeters long at the end of larval life, managed a respectable 25 kilometers, and cardinalfishes swam less than 10 kilometers before giving up.[5] Surgeonfishes, on the other hand, regularly managed 90 kilometers before giving up. That's like Michael Phelps swimming at about 6 or 7 body lengths per second for 90 kilometers—6 or 7 times his record-breaking butterfly speed, continuously, for 1,800 laps of an Olympic pool, without eating.

♦♦♦

Getting to a reef after the completion of larval life requires more than being able to swim better than Michael Phelps. The fishes also need to know where the reef is and, therefore, which way to swim. For those hatched in midwater a day or so after a spawning episode, it seems very unlikely that they have any knowledge of where they started their larval journeys. Those that spent their prehatching life in a nest on a reef might have acquired that information, but even this seems a long shot to me. Several reef scientists have been working on how late-stage larvae find their way to a reef, but it remains perplexing. On the other hand, using some amazing techniques, it has been possible to demonstrate, definitively, that reef fish larvae really do travel long distances before finding reefs and sometimes even return to where they were born. The idea that the major part of this journey is an active migration by the fish from open ocean to a reef is gaining wide acceptance; how they do it remains a burning question.

Let's consider the ways in which a larval fish might be able to detect the presence of a reef. This task is a challenge even for us if we are traveling without reliable maps—the situation that James Cook was in during his memorable voyage through the Great Barrier Reef. Reefs occur very unexpectedly, to the great detriment of sailors who think they are in deep water. How can a larval fish, out in the vast ocean, detect such structures and head toward them? A reef is visible from as much as 30 meters away in clear water; it can be heard from further than that; it can be smelled. A reef also disrupts the flow of a current and that turbulence might signal its presence, but even though fishes have impressively sensitive pressure detectors (the lateral line receptors), it's difficult to imagine a larval fish being able to resolve a turbulent pressure field when all its receptors are arrayed along its very short body. So do larval fish detect reefs and head toward them, or do they blunder into them as they swim to and fro across the ocean?

These are larval animals. They are changing continually. They lack most sense organs in their earliest days but build their sensory capabilities during the larval phase. What a fish can detect halfway through its larval life will be quite different to what it can detect toward the end of that life. In addition, relatively little work has been done on senses fish might use to detect reefs, partly because it remains very difficult to work with larval fish; rearing has proved a major challenge, and field-captured late-stage larvae start changing into benthic animals almost as soon as they are caught. In one misguided attempt to look at depth preferences of late-stage larval manini, I tried collecting fish over the shallow reef flat at night as they were coming in to settle. Silvery and translucent, they were easy to see in the light of a pressure lantern and could be easily dip-netted. My expeditions also secured the occasional octopus on nights when I had a spear with me—graduate student life is always a time for opportunism in foraging for food, and fresh octopus is delicious. But late-stage larval manini, placed in a bucket and quickly transported back to campus, inevitably swam straight to the bottom of the 2-meter tall, 20-centimeter-diameter Plexiglas tube I had waiting for them. They were no longer creatures of open water, and although they had yet to metamorphose, they were already behaving like juveniles. Jeff Leis, Dave Bellwood, and some others have been a lot more successful than I was.[6]

Nocturnal experiments using large fine-mesh cages moored in deep water near Lizard Island revealed to Dave Bellwood and Ilona Stobutzki that several damselfishes reliably swam toward distant reefs at night. Jeff has followed late-stage larvae released in the open ocean to record the direction in which a solitary swimming larva swims from minute to minute. He found that late-stage larvae frequently showed a high degree of directionality. They were swimming in a set direction, and sometimes the direction made sense. In one set of experiments larval butterflyfishes chose to swim southeast when they were released a kilometer southeast of Lizard Island reef and northwest when released a kilometer or so northwest of that reef. That is, they were swimming away from the reef that day. A damselfish tested at the same time swam southeast regardless of which side of the reef it was released on—it was headed southeast, and to hell with the reef. Such experiments show that fish can maintain direction and that a reef can influence that direction, but they don't tell us what sensory information the fish are using to do this. (Jeff favors the idea that they can hear the reef.)

A separate team, working at One Tree Reef, has demonstrated the importance of the smell of reef waters for a cardinalfish.[7] Cardinalfishes are small, colorful, nocturnally active reef fishes that brood the young in the mouth until hatching. The hatched larvae swim out of the parent's mouth and are carried out to sea by the tide for a relatively short, twenty-day larval life. Mike Kingsford, yet another reef fish biologist on the faculty of James Cook University, working with students and his sensory physiologist colleague Jelle Atema, from Boston University, had found, to his surprise, that there were genetic differences among the populations of a common cardinalfish, *Ostorhinchus doederleini*, living at One Tree Reef, Sykes Reef, and Heron Reef, three reefs with adjacent edges just 5–10 kilometers apart (see the map in chapter 3). Mike already knew, from prior studies, that tidal currents were strong enough to mix the water from these three reefs together but not to move the water rapidly out of the vicinity. Passively drifting particles (or passive larvae) would tend to remain within the vicinity of the three reefs over a twenty-day period rather than be carried farther afield. But how are genetic differences developed or retained among populations that are being mixed together every generation during larval life? What on earth is going on?

Their previous genetic studies had also shown Mike and Jelle that a common damselfish, *Pomacentrus coelestis*, also with about a twenty-day larval life, did not show such genetic differentiation among the populations on the three reefs, while the very unusual damselfish *Acanthochromis polyacanthus*, one of that handful of species worldwide that lacks a pelagic larva stage, showed clear genetic differences among the populations on the three reefs. Damselfish all deposit their eggs into nests defended by the male parent until hatching. The hatched larvae (which usually time their hatching to the peak high spring tides) get whisked out to sea by the ebbing tide, except in *Acanthochromis*, where they remain in the nest like so many freshwater sunfish or bass, cared for by both parents for about three weeks, before drifting off to their juvenile lives on the reef. Mike and Jelle thought for a while, no doubt over a few beers watching the sun set over One Tree Reef, and developed a shared hunch—a scientific hypothesis. It went something like this: Larvae of all these species are exposed to the unique smell or taste of the water on their home reefs at the time of hatching. Larvae of *Ostorhinchus* remember this smell or taste during their larval lives, during which the water they are in gets mixed with that from the other reefs. They seek out water with this flavor when it is time to find a reef to live on. Larvae of the two damselfishes either forget or do not bother to make use of this information. As a consequence, larval *Ostorhinchus* tend to head back to their home reefs, larval *Pomacentrus* find reefs, perhaps still using a more general reef smell or taste, but become mixed among all three reefs, and larval *Acanthochromis* never went out to sea anyway.

Great hypothesis, but how to test it? In a wonderful example of how to take laboratory science into the field, they did the obvious test—give fish, collected using light traps just before settling at a reef, a choice in a Y-maze between water from two of the reefs and see if they choose the reef they were collected near. Sounds simple, but it required getting samples of reef water from all three reefs and having the Y-maze ready as soon as the settling fish were caught. To do this they ended up running the Y-maze trials on a platform set up in the research vessel (a 6-meter-long skiff). In their report, they make it sound simple, but I know what it is like to bounce around in a small boat off a coral reef: this was an achievement (as are all those other studies with light traps and cages and other bits and pieces of paraphernalia taken to sea in small boats to do science).

What did Mike Kingsford, Jelle Atema, and their students find? Their cardinalfishes showed a strong preference for the waters from the reef where they had been collected; so did individuals of several other cardinalfish species. Their damselfish also showed a marked preference, as did several other damselfish species. So it's clear that water from nearby coral reefs can smell or taste differently to larval fishes. It's also clear that late-stage larvae or juveniles of several such species can distinguish water from different reefs and prefer the water from the reef they were settling into when collected. But it also seems very likely that *Ostorhinchus doederleini* uses this information to help it find its way back to its home reef at the end of larval life, whereas *Pomacentrus coelestis* does not bother. Just because I can tell one reef from another does not mean I have to use that knowledge and struggle to get back to my home reef. Yes, reef fishes are talented; no, they do not all behave in the same way, and it's fun to explore reef systems because of this.

♦♦♦

If getting to a reef at the end of larval life is an amazing feat (as it is), some reef fishes display an uncanny ability to get back to their home reef. This is a story about some amazing fishes, but also about an enduring collaboration among three reef scientists with quite different expertise. They've involved numerous students and postdocs in a long-term set of sophisticated field and laboratory studies over the past twenty years. It's a tale about reefs, but also about the exhilaration of doing science well. Their excellence raised the level of the game for other groups, including my own, and our knowledge of reef fish ecology is better because of that.

The lead actors (human) are Geoff Jones, Serge Planes, and Simon Thorrold. Geoff, educated in New Zealand, joined my lab as a postdoc in the early 1980s and subsequently, after a brief stint back in New Zealand, became one of the fish ecologists on the faculty of James Cook University in northern Queensland. He's a quiet, hard-working scientist with a brilliant mind who is far more at home in the water than on land. I think he may have gills behind his ears. Serge is a French reef scientist, a product of Université de Perpignan in southwestern France and CRIOBE (Centre de Recherches Insulaires et Observatoire de l'Environnement), a wonderful research station on the island of Mo'orea. His expertise is in the population genetics of reef fishes. Serge, who now runs CRIOBE, is an outgoing, laid-back guy with the best job in the world: he spends his time in the south of France and in French Polynesia (when he is not in the field elsewhere). Simon, a transplanted Australian, gained his Ph.D. at James Cook University before heading off to the United States and the Woods Hole Oceanographic Institution, where he has pioneered life history studies of fish using chemical information encoded in structures of their inner ears. I know all three well, and often wish I could have joined their team. Time can be a tough master.

The lead actors (fish) are a damselfish, *Pomacentrus amboinensis*, several butterflyfishes, genus *Chaetodon*, the widespread grouper known as the coral trout, *Plectropomus leopardus*, and Nemo—several different species of anemonefish, genus *Amphiprion*. In the movie, Nemo gets lost but eventually finds his way home. In real life anemonefishes are relatively uncommon, usually found living with anemones, colorful, and ideal poster children for the science, at a time when everybody aged four to ninety-four was watching the movie.

The story begins in the late 1990s, when Geoff Jones decided that if we wanted to find out where larval reef fishes go, we should tag them as they leave the reef and then watch for tagged juvenile fishes appearing on reefs. Great idea! But how do you tag a newly hatched larval fish?

Fortunately for Geoff, fish possess otoliths, minute calcified structures that float more or less freely within the inner ear.[8] Otoliths are built up by the secretion of calcium carbonate ($CaCO_3$) in a series of daily layers, like successive layers of an onion. Although otoliths are mostly of $CaCO_3$ and an organic matrix, the occasional molecule of other trace elements present in the fish and its environment gets incorporated, more or less by accident, as the otolith is assembled. Reef fish ecologists had already discovered that it was possible by counting successive rings, or annuli, to learn the age, in days, of a young fish. They also knew that settlement to a reef usually led to a characteristic deformation of the pattern of $CaCO_3$ deposition. This made it possible to determine the length of the larval life, as well as the date at which a fish settled on a reef. Also, it's widely known that the antibiotic tetracycline gets incorporated into $CaCO_3$ structures such as bones or otoliths being actively deposited and makes those structures fluorescent (bright yellow). With some pilot studies completed, Geoff knew that if he placed eggs of a damselfish into a solution of seawater containing tetracycline for an hour, those eggs would subsequently hatch into larvae indelibly marked by yellow fluorescence in their otoliths and various bones.[9] This was how he would tag larval fish before they had left the reef for their pelagic journey.

With a team of students assembled, Geoff Jones initiated his experiment. At six stretches of reef around Lizard Island, over three months in late 1994, they attempted to locate every breeding-age male *Pomacentrus amboinensis*, a common, 10-centimeter-long, drably yellow, demersal damselfish. Each of these fish was offered a PVC tile as a nest site, and all fish began defending these tiles instead of natural sites, courting females to spawn on them. Tiles, with eggs attached, were then enclosed briefly within plastic bags with tetracycline added to the water. Mostly the males put up with this disruption and resumed caring for the eggs once the bag was removed. Over the three-month period, Geoff estimates that they tagged about ten million embryos, a huge number but only about 1 percent of the larvae produced at Lizard Island reefs that season.

Eggs of *P. amboinensis* hatch in four to five days, and larval life is eighteen to twenty-one days long. So, beginning three weeks or so after they had started tagging clutches, the team began collecting settling, late-stage larvae using light traps, obtaining more than seven thousand fish that were hatched during their tagging operation. Then began the tedious job of extracting minute otoliths from the inner ears of these tiny fish. They successfully extracted otoliths from five thousand of the collected recruits, viewed them under a fluorescence microscope, and found the bright yellow signal in fifteen individuals.

I recently asked, and Geoff confirmed that they "cracked a bottle of bubbly" when they found the first fluorescent otolith. Not long after that minimilestone, Geoff appeared at an international conference. His talk included a slide that proudly said "N = 1" (meaning that he was telling a tale based on a sample size of one fish). His audience was not entirely convinced by his story. But fifteen tagged fish validate the story.

This was the first time anyone had demonstrated unequivocally that at least some reef fish return to their home reef at the end of larval life. Based on the proportion of all larvae produced that they had tagged (~1 percent), and the proportion of collected recruits that were tagged (15/5000 = 0.3 percent), Geoff was able to conclude that somewhere between 15 percent and 60 percent of larvae produced on Lizard Island reefs had returned and settled to Lizard Island reefs that season. An immense amount of work, but an important result.[10]

It was about then that Geoff Jones teamed up with Simon Thorrold and Serge Planes. They wanted to find a more easily delivered tag or a natural tag that could reliably identify fish after larval life was completed. A less risky technique. Simon came up with barium labeling of the otolith, not by somehow injecting barium into the egg, but by injecting barium into potential mothers. One injection could put sufficient barium into a female fish that she would label each of her offspring that spawning season. Serge came up with some very fancy population genetics. By collecting small fin clips from adults of both sexes in a local population of fish, he could create a genetic database against which the genetics of individual newly settled fish could be tested. But Serge did not simply test if the recruit had been spawned within that population. Using a technique called parentage analysis, he was able to test each individual juvenile against the set of adult males and females to identify its most likely parents. This is like using human DNA as evidence in a paternity suit.

Collectively, Geoff, Simon, and Serge decided to start by applying these new tagging techniques to relatively uncommon fish species. In that way they could have a reasonable chance of sampling most of the adults living on any stretch of reef.[11] Their first test subject . . . Nemo.

Now it so happened that in the mid-1990s Geoff Jones had a graduate student who had previously been the dive manager at a tiny place called Walindi Plantation Resort in New Britain, Papua New Guinea. The owner of the resort was a bit of a visionary, interested in sustaining the environment (such people really do exist), and had established a nongovernmental organization (NGO) called Mahonia na Dari (Guardian of the Sea). The plan was for the NGO to operate a research and conservation center. He was seeking researchers who'd be interested in contributing their expertise while doing research there.[12] In 1996, Geoff visited and fell in love with the New Britain environment, and particularly the large Kimbe Bay, which stretched out in front of the resort. Pure serendipity, but that's how it came to pass that Geoff, Serge, and Simon chose sites in Kimbe Bay, on the northern coast of New Britain, for their field studies. Geoff told me in 2018 that he has gone to Kimbe Bay at least twice a year ever since.

Their first target was the panda clownfish, *Amphiprion polymnus*, also known as the saddleback clownfish, a gorgeous, reddish brown to black clownfish with two (sometimes three) white saddles. Geoff had stumbled across a group of them in anemones around tiny Schumann Island—a small, vegetated island scarcely 250 meters in diameter, near the western shore of Kimbe Bay—but had seen the species nowhere else in the bay. During three-month periods in 2002 and 2003, the team applied Geoff's tetracycline-labeling trick to nests of eggs at the bases of anemones in the shallows around Schumann Island. Whereas Geoff's team had marked 1,926 clutches of *P. amboinensis* eggs at Lizard Island and still tagged less than 2 percent of all eggs produced, at Schumann Island, only thirty-three pairs of clownfish were present (occupying thirty-three of the forty anemone clusters there). They tagged every egg produced during the three-month periods and then collected all sixty-three recruits in 2002 and all seventy-three recruits in 2003. Clownfishes recruit only to anemones, so they are reasonably certain they had collected all recruits to Schumann Island (other than ones that were taken by predators soon after settling).

I remember Geoff Jones commenting, before he did this experiment, that he wanted to select a rare species for his studies because it would be easier to sample the population adequately. I also recall thinking how unlikely he would be to get any results if he worked with a rare species. I was very wrong! Ten of the recruits in 2002 and twenty-three of those in 2003 (16 percent and 32 percent of all larvae recruiting, respectively) were tagged! These animals had spent their nine-to-twelve-day larval lives at sea and had then returned to a tiny, 2-hectare stretch of reef surrounding an even tinier island. In 2003, the team collected fin clips from all breeding adults and genotyped all collected recruits, using eleven microsatellite DNA markers. In Serge Planes's lab, these genetic data were analyzed for parentage of the recruits to deliver a comparable result—32 percent of the recruits had parents among the breeders at the island.[13] Both methods had provided the same surprising result.

The majority of larvae obviously disperse more widely because the 68 percent of recruits that were not from Schumann Island traveled a substantial distance to settle there (Geoff's team hasn't found any panda clownfishes anywhere else within Kimbe Bay). But that almost a third of arriving juveniles were returning home is surprising. That fact also begs another question: Why is the panda clownfish, and the two species of anemone it occupies, so rare elsewhere in Kimbe Bay? Nobody knows.[14]

The Schumann Island study was just the start for Kimbe Bay. In May 2007, postdoctoral student Glenn Almany published his results in *Science*, along with Thorrold, Jones, Planes, and Michael Berumen, another member of the growing team. Once again they did the fieldwork in Kimbe Bay, but now using Simon Thorrold's new tagging technique using barium. This time, Almany and friends compared the larval recruitment of another clownfish, *Amphiprion percula* (somewhat more common than *A. polymnus*, often referred to as *the* clownfish, and the same species that stars in *Finding Nemo*), and a common, 15–20-centimeter-long butterflyfish, *Chaetodon vagabundus*, at tiny Kimbe Island slightly west of the middle of Kimbe Bay and far from any other reefs (fig. 20). While *A. percula*, like *A. polymnus*, has demersal eggs and a brief larval phase lasting approximately eleven days, *C. vagabundus* has midwater spawning and a more typical reef fish larval duration of about thirty-eight days. I would have expected substantial differences among these species in the pattern of larval return. I was wrong again!

Geoff's tetracycline tagging method could work only for fish with demersal nests. Simon's new technique tagged the females by injecting them with a barium chloride solution that had been highly enriched with ^{137}Ba and depleted in ^{135}Ba compared to naturally occurring barium compounds. Dosed with barium chloride, the females proceeded to label all their eggs, and the developing larvae grew otoliths with a dense accumulation of this barium at the core. Prior laboratory tests had confirmed that females would survive this treatment and lay labeled eggs for at least one full spawning season. This method would work just as well on midwater spawning butterflyfishes as on clownfishes with demersal nests. I'm making it sound simple,

Figure 20. Nemo, *Amphiprion percula,* safe at home in her anemone on a reef in Raja Ampat, Indonesia. This fish has demonstrated a surprising capability to return home following its short larval life. Photo © A. J. Hooten.

but there was considerable insight involved in dreaming up this way of tagging fish. To read the tags, the team had to extract otoliths from recently settled juveniles and vaporize them using an ICP-MS, an inductively coupled plasma mass spectrometer. This instrument and the lab facilities to sustain it has a cost comparable to several years' worth of my entire research program, a sign of how complex and expensive modern coral reef ecology has become. But the data were definitely worth it!

Kimbe Island, an islet even smaller than Schumann Island, together with its own tiny surrounding reef, is a 30-hectare blip on top of a pinnacle that rises from deep water in the middle of Kimbe Bay, and 10 kilometers away from the nearest other reef. If Geoff Jones's results at Lizard Island had been surprising, and if the subsequent results at Schumann Island were amazing, the results from Kimbe Island were simply mind-blowing. Glenn collected fifteen newly settled juvenile clownfishes and seventy-seven newly settled butterflyfishes; nine clownfishes and eight butterflyfishes were found to carry the distinctive ^{137}Ba label, leading to a conclusion that about 60 percent of larvae of both species had managed to get back home to Kimbe Island![15] I well remember when I first read their paper; I sat there stunned. We had all come to believe that reef fish larvae could actively find reefs, but none of us expected that they would be that effective in getting home.

The studies by Geoff, Simon, and Serge at Kimbe Bay have continued. An article in volume 1 of the new journal *Nature Ecology and Evolution* appeared in 2017.[16] This time Glenn Almany led a team of sixteen coauthors, including the three leaders, to report results of a massive extension of the work on *A. percula* and *C. vagabundus.*

In an enormous fin-clipping effort, they collected genetic samples from 2,546 adult *A. percula* and 2,021 adult *C. vagabundus* in 2009, and 2,913 and 4,858 adults of the same species in 2011. These samples came from eight locations scattered around Kimbe Bay and as much as 120 kilometers apart. Their sampling included almost all adult pairs of clownfishes and between 12 percent and 77 percent of adult butterflyfishes at each site. They then returned to each site and collected juveniles that would have settled during the six months following their initial fin-clipping exercise. Over the two years they caught 2,994 clownfishes and 1,943 butterflyfishes. In the paper, they name and thank a team of twenty-one volunteers who helped in these massive collecting efforts.[17]

They used the adult fin clips to build a genetic database of breeding adults against which they attempted to match each collected juvenile. They were able to identify one or both parents of 844, or 28 percent of all juvenile A. *percula* collected. They also identified one or both parents of 89, or about 4.5 percent of juvenile C. *vagabundus* collected. The lower success with the butterflyfish is partly due to the less intensive sampling of adults of this species, but also a sign that this species disperses more widely than does the clownfish. These fish for which one or more parents were known provided definitive information on site of origin and destination at the end of larval life, and the team used these data to compile what are called dispersal kernels for each species.

Dispersal kernels are graphs showing the pattern of settlement with distance from the source location. They permit estimates of the average extent of larval dispersal, an important clue to the spatial scale at which reef fish larvae live their lives and the scale at which local populations are connected to one another. The data for A. *percula* suggested that the average dispersal distance was about 15 kilometers, with 90 percent of larvae settling within 30–40 kilometers of where they were spawned. For this species, the populations within Kimbe Bay form a well-connected metapopulation, relatively separated from populations farther west or east.[18]

Chaetodon vagabundus gave less clear results; the average observed dispersal distance was about 50 kilometers, but the calculated dispersal kernel did not differ significantly from a horizontal line, suggesting an equal probability of settlement at all distances from the source. Despite extensive sampling across Kimbe Bay, the scientists had not sampled sufficiently broadly to have captured the limits of dispersal, and a realistic assumption is that populations of this species within Kimbe Bay are well connected through dispersal with populations outside the bay and along the north coast of New Britain.

Their results also revealed considerable differences from place to place, and from year to year, in the degree of self-recruitment (the tendency of animals produced at a site to return home at the end of larval life). The temporal variability was quite high at Kimbe Island, for which they now had data from 2005, 2007, 2009, and 2011. For C. *vagabundus*, self-recruitment at Kimbe Island had been about 60 percent in 2005 but only around 5 percent in each of the other three years. This pattern of high variability in time and space has been seen in all aspects of larval dispersal and settlement in all studies of reef fish around the world. I've even argued that the one thing that is dependable about these processes is that they are hugely variable! The variability tells me that it will be very difficult to define dispersal kernels for most reef fishes in most places, but it also indicates that the path taken by an individual larva from hatching to settlement must be strongly affected by such physical forces as highly variable current patterns, as well as by the capabilities of the larvae. Reef fish larvae are amazing creatures; they must be quite clever at finding reefs and at getting back home to the reef where they were spawned, but they live in a challenging environment and are just doing the best that they can. Actually getting home is for most an unlikely outcome.[19]

The study of larval dispersal has been one of those long-standing challenges for those of us interested in the ecology of reef fishes. We've made progress because so many different approaches have been drawn in to tackle this challenge. I think we have come a very long way from where we were in the mid-1950s, when Jack Randall wrote, "On two occasions before midnight at low tide in ankle-deep water along the shore of the Ala Wai Yacht Basin near the entrance to the Ala Wai Canal, manini [larvae] were observed just coming into the area from deeper water. They did not passively float into the shallow zone but swam in rapidly. One crossed several times through the beam of a head lamp before it was caught. If it is assumed that these fish were in deep water beyond the breaker zone before nightfall, then they must have actively swum into the harbor area and not been carried in by any tidal currents. Prior to low tide, tidal currents would be flowing out of the yacht basin and not into it."[20]

Jack was correct; they were actively swimming in toward the shallows where manini spend their juvenile lives. Now we know that the vast majority of reef fishes actively choose their settlement locations and that at least some can find their way back to the site where they were spawned. We know they are very capable swimmers, at least in the later stages of larval life. We know that they respond to the presence of reefs while at sea and that they can hear and smell reefs. We have no idea how homing (as opposed to just finding reefs) is accomplished, although the work of Mike Kingsford and Jelle Atema suggests that they may use olfaction much as homing salmon do in streams. Nor do we yet have much information on when in the larval life the animal becomes capable of and begins to use its capacity to find reefs. And of course we remain in the dark about why they do all this.

♦♦♦

Early in this chapter, I mentioned that there were half a dozen reef fish species that do not have pelagic larvae. *Acanthochromis polyacanthus*, for example, is an abundant, dappled gray, 12-centimeter-long damselfish found throughout the Great Barrier Reef and beyond and is quite conspicuous about caring for its offspring. In doing so, it behaves as if it has been taught parental care by a North American freshwater bass or sunfish (family Centrarchidae) or one of the myriad species of cichlid fishes (Cichlidae) found in the African Great Lakes. The parents share care of the eggs and young in a shallow nest, allowing the young to feed off mucus on their flanks or shelter briefly within their mouths, until the fry are large enough to fend for themselves. Because damselfishes have been so well studied, it is reasonable to state that this is the only damselfish that does not have a pelagic larval stage, but this raises a vexing question: Why don't all the other damselfishes extend their care of their eggs into care of the young?[21] Parental care of young fry is clearly an evolutionary possibility for damselfishes, or *Acanthochromis* would not exist. Surely, by caring for the young, individuals of other species would enhance their fitness. Letting the offspring disperse to the open ocean seems foolish. The open ocean is not a safe place, nor is it a place producing ample food. In fact, the tropical ocean is pretty much a desert. About the only argument in favor of pelagic larvae that makes sense is that by dispersing offspring to the open ocean, a reef fish increases its chance of success by spreading its offspring out across many different reef locations. This could be true if coral reefs are actually rather risky environments in which to live. Perhaps that is the case—we know they are risky on longer time frames of hundreds or thousands of years because reefs come and go.

This question, like all good "why" questions, cannot be answered definitively using science. There is no way to test the hypothesis that pelagic larvae are necessary because coral reefs are risky places to live. We can declare this augustly; we can build models that simulate it; we can pound on desks should anyone oppose us. But there is no way to prove it, and all the time, *Acanthochromis* and those other nonconformers are there to mock us—they are doing just fine without pelagic larvae. But we sure can tackle the how, when, where, and what with questions to convince ourselves that we are finding out deep truths. Doing science on coral reefs is just plain fun, and reefs are wonderful at throwing questions out to scientists.

One more thing, let's also remember that it is not just the reef fish that appear to treat reefs as risky places to live. The overwhelming majority of reef creatures have pelagic larvae. For some sea squirts (subphylum Tunicata), small, somewhat spongelike, filter-feeding creatures closely related to vertebrates like us (yes, we have some modest relatives in our family tree), the larval phase is only minutes to hours long, and it proved possible to follow newly hatched larvae in the water column until they settled to the substratum.[22] However, the great majority of marine invertebrates have larval phases twenty or more days long, and some crustaceans, such as the spiny lobsters, have larval phases approaching twelve months long. In other words, they are much like fish. One big difference between the invertebrates and fishes has been that fish have offered us more ways to learn about their larval lives. Their otoliths provide a daily record of growth, and a mark defining the settlement day, while also incorporating strange chemicals as they grow, thereby facilitating tagging. Larval fish seem behaviorally more capable than many invertebrate larvae, although surprises are probably waiting for us. I for one will not be surprised when scientists discover that some mollusks or crustaceans also have larvae capable of navigating back toward home after a larval life at sea. Reefs are filled with incredibly talented creatures.

EIGHT

Wondrous Reef Chemistry

I have an aversion to chemistry. As an undergraduate in zoology at the University of Toronto, I was required to take organic chemistry, a year-long course with three lectures and a three-hour lab session each week. The lab sessions were held on Saturday mornings. Even worse, I am particularly sensitive to volatile organic compounds. In those days, teaching labs in chemistry departments were built with little regard to air quality. I'm not even sure if there was a single laminar flow hood or even an exhaust fan over any workbench in our lab. As a consequence, I went home with a migraine almost every Saturday of that miserable year. Sure kept my interest in partying under control!

It's perhaps not surprising that although I passed the course, I studiously avoided all matters chemical from then on. And as with muscle, if you do not use knowledge, you lose it. Fast forward forty-five years to a time when ocean acidification was becoming a growing concern, and to a discussion among collaborators working on a manuscript. The complexity of chemical interactions of carbon dioxide, water, and calcium ions in the ocean, being dealt with quite handily by most of my ecologist colleagues, left me feeling that I had stepped into the wrong conversation by mistake and was listening to classical musicians discussing the intricacies of twelve-tone serialism in Mandarin. While I learned that surface waters with an aragonite saturation (written $\Omega_{aragonite}$) less than 3.3 represented conditions in which coral reefs would no longer be growing, because rates of erosion and dissolution would exceed rates of calcification, I had to take this statement at face value since I lacked the chemistry foundation to understand it properly. (Nevertheless, I proudly kept my name among the authors of the eventual paper, realizing that there were lots of other things in it to which I had contributed and did understand thoroughly. Such is interdisciplinary collaboration!)[1]

This story is my mea culpa, because in this chapter I explore some of the many ways coral reefs exhibit wondrousness in the handling of nutrients and energy to build their richness and complexity despite living mostly in a desertlike, nearly sterile environment. Life anywhere is fundamentally a process of building and then breaking down organic molecules. It's chemistry, and when it comes to chemistry, I may not be a reliable guide. These energy-rich molecules are built using energy of sunlight in photosynthesis, and they are broken down to capture the stored energy needed to drive the metabolism of all cells. The molecule-building part of the process is called primary production, and the breaking down is sometimes called respiration. On a reef, calcification is an essential energy-using process engaged in by many species, and part of community respiration.

♦♦♦

The sheer abundance and diversity of coral reefs are mirrored by their unrivaled rate of productivity among marine ecosystems. Coral reefs are the most productive ecosystems in the ocean, in terms of gross primary production—the production of organic matter through photosynthesis. Yet the waters of the tropical ocean, especially far from continental shores with their supplies of nutrient-laden runoff, contain very low levels of nutrients. How is such production possible? Paradoxically, the most productive marine ecosystem exists in a veritable desert! And some of the first quantitative ecology done on coral reefs tackled this paradox.

To understand the puzzle of reef productivity, we must take another look at the corals themselves and at their algal symbionts. In earlier times people found corals quite mystifying. Aristotle's pupil Theophrastus, around 300 BCE, included *kouralion* in his treatise on minerals, thus giving us the word *coral*, and described red coral as "similar to a stone, is shaped like a root, and found in the sea."[2] They were puzzling rocks, or were they plants? That they are indeed creatures becomes apparent if they are removed from the water:

first they secrete a lot of mucus; then, over the next couple of days they smell really bad as they decompose. Rocks don't do that. Yet even in death, the complex structure of corals is largely retained, although it is brittle and easily reduced to rubble.

By Pliny's time (23–79 CE) coral had progressed from being a rock to being definitely thought a plant. Linnaeus classified corals as "lithophytes," stony plants, in the first edition of his *Systema Naturae* in 1735. There coral floundered until 1752, when the French scientist Jean-André Peyssonnel read a paper to the Royal Society of London, in which he provided evidence that corals were colonies of minute animals. Linnaeus reached the same conclusion. In 1758 his tenth and definitive edition of *Systema Naturae* placed corals as animals within the order Anthozoa (meaning flower animal), where they have remained, more or less, until today.

By Darwin's time, one hundred years later, science broadly accepted that corals were animals; Darwin called them "polypifers," signifying that they were colonial organisms comprised of minute "polyps." But what kind of animal? By 1883, the tiny yellowish objects visible within coral tissues had been identified as symbiotic, single-celled algae that lived within coral cells; they'd been named zooxanthellae (meaning yellow things within animals).[3] That led to much conjecture and experimentation to determine whether these symbionts were parasites or commensals, whether the coral fed on this internal garden of plants, whether the coral provided nutrients to the plants, or whether the algae provided nutrients or energy to the corals without being eaten. The actual relationship, physiological and ecological, between the coral and their symbiotic algae became one of the first major questions tackled when biologists began serious research on coral reefs early in the twentieth century. (A second was the question of how coral reefs were formed and whether Darwin's subsistence theory of reef formation was correct. Third in this list of questions was simply how to describe the structure and organization of this complex ecosystem so seemingly unlike any other.)

By the 1920s, it was generally agreed that these symbiotic algae were a type of dinoflagellate, a single-celled alga usually having two long whiplike flagella, one projecting forward and the other lying within a groove around its middle.[4] As well as occurring in corals, they occur in anemones, soft corals, certain sponges, giant clams, and some other reef creatures.

In 1928, the world's first major effort to explore the science of coral reefs occurred. This was the yearlong Royal Society Great Barrier Reef Expedition to Low Isles on the central Great Barrier Reef.[5] The interaction of corals and their symbionts was a central topic being investigated. Among other discoveries, Maurice Yonge, twenty-seven years old and expedition leader, reported that corals release oxygen to the water in daylight and that starved corals, or corals kept in the dark, would lose their algal symbionts. In the early 1940s, the Japanese biologist Siro Kawaguti, in pioneering pre–World War II work in Palau, had placed these symbiotic algae within genus *Gymnodinium* and shown that calcification by (and therefore growth of) the coral was strongly dependent on their presence. The ground was now prepared; World War II ironically provided for the first serious studies of reef productivity, although understanding of calcification had to await new molecular approaches to biology.

After World War II, research on coral reefs was substantially assisted by the money being spent by the U.S. Atomic Energy Commission (AEC) as the race to test atomic weapons heated up. For some reason, the United States, which had been testing prototype bombs in the deserts north of Alamogordo, New Mexico, decided it should use atolls in the South Pacific as suitable locations for further hydrogen bomb testing![6] Because one of the more pressing questions concerned the environmental effects of radiation following detonation, the AEC ensured that the scientific teams included environmental scientists with appropriate skills. These scientists came largely from the university sector. Obviously, they first needed to collect baseline data with which to compare data gathered after a test, and ecologists, being ecologists, made sure that the collection of baseline data became useful in advancing coral reef science.

In 1946, Marston Sargent was just out of the U.S. Navy and a new faculty member at Scripps Institution of Oceanography, on the coast at La Jolla, California. He and fellow Scripps scientist Tom Austin soon found themselves at Rongelap (an atoll east of Bikini Atoll in the northern Marshall Islands) with funding from the AEC. They were part of a large team charged with producing baseline environmental data in preparation for the nuclear testing program to take place at nearby Bikini.

In planning their work, they cleverly took advantage of a physical feature common on atolls: between the islets that typically surround the lagoon, and especially on the windward side, extensive reef flats form a low-relief shallow bench between the open ocean and the lagoon. Water flows linearly across these benches from open ocean to lagoon although varying in velocity with tidal phase. Physiologists had developed the flow cytometry technique to look at metabolic processes in tissues or organisms living in an aquatic medium in the lab, and Sargent and Austin used the same approach, minus the clean lab bench and glassware, to examine the metabolic activity of the community of organisms living on the bench.[7]

By measuring the difference in oxygen concentration in the water between the upstream (outside) and downstream (lagoonal) edge of the bench, and the rate of flow across the bench, they were able to compute the rate of production (or consumption) of oxygen as each parcel of water passed over the reef environment.[8] They found that the water was saturated in dissolved oxygen at the seaward side of the bench but that it *increased* in oxygen concentration in the few minutes before it reached the lagoon edge. Photosynthesis was generating sufficient oxygen to provide a surplus beyond that used by respiration of plants and animals living on the bench. During mid-daylight hours, this production of excess oxygen was at a peak; the rate of production declined from there until evening, and during the night there was a decrease in oxygen content across the bench (because organisms continued to respire but no photosynthesis was happening). Looking at the pattern through twenty-four hours, and visualizing the bench as a series of adjacent centimeter-wide strips each extending from ocean to lagoon, they were able to compute the total excess oxygen produced per day as 3,000 milliliters of oxygen per centimeter of reef bench surface (14,000 milliliters excess during daylight minus 11,000 milliliters deficit during the night). That 3,000 milliliters of oxygen is equivalent to the incorporation of 1,600 milligrams of carbon into organic matter per centimeter of reef bench. On an annual basis that would be 190 grams carbon per square meter per year.[9]

This is the estimated net primary production (production in excess of consumption in respiration); the reef was producing an annual surplus of organic matter. Using similar logic, Sargent and Austin computed the gross primary production (total photosynthetic production) to be 1,500 grams carbon per square meter per year, a rate many times higher than in the open ocean or in the lagoon.

A far more influential paper, telling essentially the same story, appeared a year later, based on fieldwork done in 1954.[10] Eugene P. Odum and Howard T. Odum were brothers and prominent American ecologists of the 1950s. By 1950, Eugene was well established at the University of Georgia and had successfully tapped into the AEC for sustained major research funding that established what became the University of Georgia's Savannah River Ecology Laboratory.

In the summer of 1954, Howard was between jobs when Eugene secured another major AEC grant, this time for pioneering work on the ecology of Enewetak Atoll in the northern Marshall Islands west of Bikini.[11] Atomic testing at Enewetak had commenced in 1948 with Operation Sandstone, and the first large test of a thermonuclear device, Mike (10.4 megatons, or Mt), had taken place in October 1952.[12] A smaller (1.6 Mt) test occurred in May 1954, a few weeks before the Odums arrived (they did their research upstream and upwind of the contaminated northwest side of the atoll). There were eleven more tests in 1956 and twenty-two in 1958. Subsequently the lagoon became a target and recovery site for long-range missiles from California. Ultimately, the atoll was decontaminated and inhabitants were encouraged to return, although parts of the northern atoll rim still remain too radioactive for human use. As part of the mission, the AEC continued to support a small marine lab at Enewetak until 1983, when the U.S. government got tired of paying for fuel for the small generator![13]

The Odums arrived for a six-week stay at Enewetak Atoll in the summer of 1954. The fieldwork they accomplished was published in *Ecological Monographs* and subsequently won them the Mercer Prize from the Ecological Society of America as the best ecological paper published by "a younger author" in the previous two years.[14] It's amazing what the U.S. military achieves.

The article remains fascinating, especially when you realize how little was known about the ecology of coral reefs when the Odums landed on Enewetak. They had a copy of the paper published by Sargent and Austin, and they used the same method for determining "community metabolism" on one of the extensive shallow benches between islets of Enewetak, but they did a far more detailed job characterizing the shallow-water ecological communities, determining standing biomass (the total weight of living matter) of various community components, and measuring many aspects of productivity (fig. 21). They demonstrated that a typical coral head contained three times more plant tissue than animal tissue (lots of filamentous algae living within the inert dead skeleton as well as the symbiotic algae), leading to the claim by some that corals are more plant than animal. They estimated gross primary production of 24 grams of carbon per square meter per day for the 322-meter-wide reef bench, total community respiration slightly less, and export to the lagoon of organic matter, chiefly as plankton, of 0.4 grams of carbon per square meter per day. Their data suggest a gross rate of primary production twice as great as that computed at Rongelap by Sargent and Austin; reasons for this likely include a real difference between the reefs (Rongelap bench was described as barren) plus more precise techniques. Subsequent work through to the 1980s by other workers using far more sophisticated methods has revealed that these early estimates from Rongelap and Enewetak were both outliers (Rongelap low and Enewetak high); the consensus now is that gross primary production on a typical coral reef is about 7 grams of carbon per square meter per day, similar to that of typical evergreen broadleaf forest (7.5 grams of carbon per square meter per day).[15] (The rate varies from place to place and can be as high as 20 grams of carbon per square meter per day over areas of 100 percent coral/algal hard substratum.) It's also usual for the overall rate of production to approximately equal that of respiration by a reef community—meaning that the reef community is largely self-sustaining, producing the organic material it consumes in surviving.[16] Productivity of phytoplankton in the open waters near reefs is typically less than 0.1 grams of carbon per square meter per day, between one and two orders of magnitude less.

Atolls like Rongelap and Enewetak are surrounded by waters that are ex-

Figure 21. The section of shallow bench at Enewetak Atoll where the Odum brothers did their study of coral reef productivity is in the foreground, but the bench stretches into the distance in this aerial photo looking northwest. The inset satellite image of the atoll reveals a small deepwater cut in the atoll rim in the southeast that lies more or less under the camera in the main photo. The relatively sharp curve near the second island along the bench is the most easterly point on the atoll rim. Photo © Patrick L. Colin; inset derived from Landsat image courtesy NASA/USGS.

tremely poor in nutrients such as nitrogen and phosphorus (the Odums reported 0.1 grams of nitrates per cubic meter). While fringing and barrier reefs along continental margins occur in waters richer than this, an increase in nutrients (and turbidity) correlates with a reduced diversity and perhaps productivity of coral reefs. Reefs really are amazing oases in a veritable desert. So, how do they manage to be so productive in such circumstances?

The Odums suggested two possible reasons, and both have been borne out by more recent work. First, some blue-green algae and other bacteria, living among the algal turfs that cover nonliving surfaces in shallower parts of the reef, actively fix atmospheric nitrogen as nitrates.[17] Second, there is highly efficient shuffling of nutrients among producers, herbivores, and carnivores, which live in intimate proximity to one another. This keeps nutrients in play and maximizes their use by the living reef. Who are among the most important players in nutrient shuffling? Those tiny algae living within coral tissues! All of which brings me conveniently back to the relationships between corals and their symbiotic algae, relationships that are fundamental to both photosynthesis and calcification.

♦♦♦

If it took a long time to determine that zooxanthellae were symbiotic dinoflagellates living inside the cells of the coral animal, we still don't have a definitive understanding of the diversity of these symbionts or of their distribution among different species of coral. The original genus, *Gymnodinium*, got replaced in 1962 by creating a new genus, *Symbiodinium*, and, for a while, it was believed there might be only one species, *Symbiodinium microadriaticum*. Fat chance! Recent genetic and molecular research has proved that to be a gross simplification (although I'm going to keep things simple by referring to all of them simply as algae).[18]

When free-living, algal symbionts all look like a typical dinoflagellate complete with two flagella. They are tiny, ovoid cells about 5 micrometers (0.005 millimeters) in length, with a constricted "waist." One flagellum emerges near the anterior end, and the other emerges in the waist, circling the body like a loosely fastened belt. The first flagellum moves the organism forward, while the second causes it to spin on its axis.

Of course, most of the time, algal symbionts are not free-living and don't look a bit like this; they are seen, in their millions, as tiny, round, yellowish cells, packed with chloroplasts, living within the cells of corals. In this vegetative form, the cells are about 10 micrometers (0.01 millimeters) in diameter. They reproduce asexually by simple cell division to produce two daughter cells. Alternatively, they can divide to form flagellate cells that are active and free-living (and it appears that the coral can control which form of asexual reproduction its symbionts engage in). Sexual reproduction also occurs and apparently can occur within both the symbiotic and the free-living stage.

Despite the metabolic importance of algal symbionts to corals, most coral species rely on being able, at the end of larval life, to pick up free-living dinoflagellates (which then transform into vegetative cells), rather than passing these minute creatures down from mother to daughter within the egg.[19] Inside corals, algae live within the cytoplasm of specialized cells of the gastrodermis (endoderm) of the host coral in the most intimate type of symbiosis possible. They also occur as intracellular symbionts in soft corals, gorgonians, anemones, jellyfishes, flatworms, and sponges, and extracellularly in giant clams (*Tridacna*) and some other mollusks.

Small, but hugely abundant on coral reefs, hidden within the cells of their host organisms, algal symbionts play a major role in keeping the coral reef system running. Indeed, in one sense, coral reefs are ultimately the result of cellular-level processes that take place within the coral animal. These rely on intricate exchanges between various organisms including bacteria and these algae. The algae are arguably the most important of the participating organisms, perhaps as important as the corals themselves. They are the photosynthesizers, using sunlight to produce sugars, lipids, amino acids, and oxygen, which are used by the coral host in its own metabolism and provide up to 95 percent of the coral's metabolic requirements. Waste products from coral cellular respiration, especially carbon dioxide and water, are taken up by the algae and used in photosynthesis.

Advances in molecular methods are revealing lots of different types of algal symbiont, but today it is still not clear how many species should be recognized.[20] For a while, researchers spoke in terms of different clades of *Symbiodinium*, meaning, "*We have these different groups of organisms that we think might be groups of related species, but we aren't sure, so we'll just call them clades.*" As I write, eight clades have acquired names as formally described genera, and there will likely be fifteen genera in total.

The lack of a settled taxonomy for these algae, in which every organism has an agreed name and clear relationships with others, reminds me of the similar lack of settled taxonomy in other kinds of (larger) reef creatures that I experienced in the 1970s and as I discussed in chapter 4. Deciphering relationships among these tiny intracellular algae had to await sophisticated molecular techniques, but our confusion concerning their diversity is just one more example of how incredibly diverse and rich coral reef systems are.[21] Now researchers are in a real race against time to sort out the taxonomy, because how corals and their symbionts interact is crucial to how they survive climate warming. If we are ever to help them adapt as the planet warms, we need a much deeper understanding of these relationships than we now have. To do that, we have to begin by figuring out who's who.

Within geographic regions, different types of algal symbiont occur at varying depths or in different environments, and different types dominate within or are exclusive to certain species of coral. While most coral species are able to form symbioses with more than one type, corals vary in this trait from some, such as species of *Montastrea* and *Orbicella* in the Caribbean, which typically contain two or more types within single colonies, to others, the great majority, in which a single type usually dominates each colony. The details of the patterns of distribution of these algal symbionts—by region, by habitat, and by host coral species—are still being worked out, as is the overall richness of species. Understanding these patterns is important, because the different species have varying capabilities and tolerances. A coral's ability to cope with its environment depends on which types of algal symbiont it harbors.[22]

♦♦♦

It's one thing to talk about corals and their algal symbionts achieving high rates of photosynthesis and using much of the organic molecules built to provide energy for calcification. It's quite another to comprehend how they do these things. That's because photosynthesis needs a low pH environment and calcification requires a high one. Somehow the corals manage that incompatibility.

In previous chapters, I've talked about coral reefs as neighborhoods containing residents of many species that interacted in sophisticated ways with one another to create the social network that is a coral reef. But I have come to understand that coral reefs need to be looked at from many perspectives (and are usually quite wonderful from each). If we narrow our perspective down to the microscopic and peer closely inside the living coral tissue that covers coral colonies, we discover another world of neighborhoods, also occupied by many different creatures, interacting with one another in complex ways to achieve the tasks on which depends the existence of coral reefs themselves. In these neighborhoods, the species, other than corals and algae, are microbial, and the interactions are chemical and physiological rather than behavioral. These are neighborhoods that I have to work hard to understand, but they are every bit as amazing as those larger neighborhoods I understand quite well. These microneighborhoods are also ones in which the corals are far more than architecture—the interactions among corals and their symbionts are important, and integral to their ecological success. Welcome to the coral microbiome, an analog to our own human microbiome.

We'll begin with the anatomy of a coral because that provides the architecture, the neighborhoods, within which the microbiome lives (fig. 22). As discussed in chapter 2, the coral polyp is a simple organism with a single body cavity, the gastric cavity, having a single opening, the mouth, surrounded by tentacles. Its body wall has two cellular layers, the outer epidermis and the inner gastrodermis; between them is a thin, jellylike layer, the mesoglea (we and most other multicellular animals are creatures that as embryos have an ectoderm, an endoderm, and between them a mesoderm—a third cellular layer from which we build our muscles and most internal organs). The gastrodermis is folded into a series of mesenteries that extend up toward the mouth from the septa on the skeletal base (remember those septa discussed in chapter 2?), greatly expanding the surface area of the gastric cavity. That's about it—no organs, in the sense that we have kidneys, livers, and brains, although the epidermis and gastrodermis are both composed of many different types of cells specialized for different functions.

In coral colonies, the body walls of adjacent polyps are connected (epidermis, mesoglea, gastrodermis, and gastric cavity of adjacent polyps are all interconnected), although the extent of this connection between polyps varies among species. Each polyp sits within its own cup on the surface of what is an external skeletal mass of calcium carbonate rock. The portion of the epidermis in contact with the skeleton can be very extensive, given the depth of the cup and the complex architecture of the skeletal surface—all those septa demand lots of folding of epidermis. This skeleton-contacting epidermis, the calicodermis, is the coral's calcifying tissue, its skeleton-building machinery.

The coral's anatomy provides several different neighborhoods for symbionts; these offer quite distinct conditions, and they are occupied by varying mixes of symbionts. Algal symbionts are found only inside specialized cells,

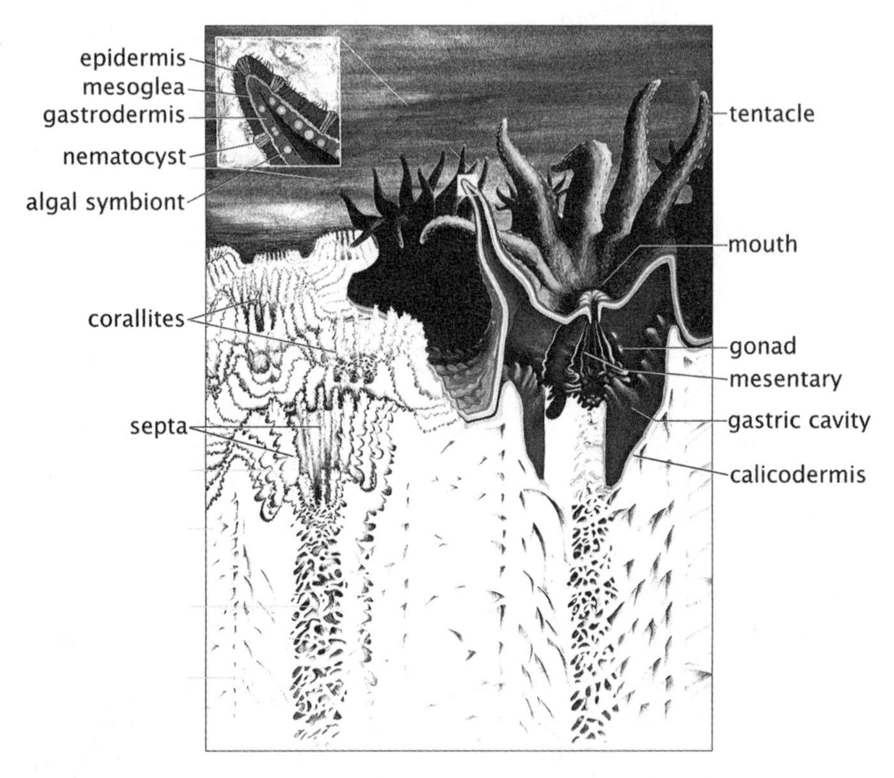

Figure 22. The anatomy of a coral is shown in this cutaway view, with the living coral tissue peeled away from the left-hand side of the carbonate skeleton and the nearest polyp sliced vertically through its center. One tentacle tip is shown enlarged, in cross-section at the upper left. Notice how the gastrodermis, which covers the entire inner surface of the polyp, is heavily folded into mesenteries, greatly expanding the surface of the gastric cavity. The mesenteries tend to align with the skeletal septa below them. The coral polyp can be expanded, protruding well beyond the outer surface of the skeleton, as shown, or retracted to protrude scarcely at all. Image modified with permission from J. E. N. Veron, *Corals of the World*, vol. 1 (Townsville, Qld.: Australian Institute of Marine Science and CRR Australia, 2000), drawing by Geoff Kelly.

the symbiocytes, in the gastrodermis. They are most abundant in the upper/outer portion of the gastrodermis (which receives the most light). Abundances can exceed a million algal cells per square centimeter of polyp gastrodermis. Numerous different bacteria are also reliably present as members of the microbiome, but given that these are far smaller than a dinoflagellate, scientists look for them using genetic methods rather than a microscope.

Consequently, we are less clear about where they reside within the coral, although it's increasingly evident that they also mostly occur in specific places. For example, there is a rich fauna of bacteria, and a few other microbes, that occur in the layer of mucus that covers the coral's external surface. A different group resides within the calicodermis, where calcification takes place. Other bacteria occupy the gastric cavity and still others the tissues of the body wall, where some are intimately associated with the algal symbionts in the symbiocytes.

Although each subset of bacterial symbionts may include pathogens (especially true of the mucus-inhabiting group), each also contains species that play positive roles in the bits of physiology/metabolism that take place at each location. For example, the mucus-dwelling "exterior" bacteria include some that produce bacteriocides that protect against foreign pathogenic forms and some that produce sunscreens protecting the coral and its symbionts from ultraviolet radiation damage. The microbiome of the gastric cavity contains a number of taxa that aid in the digestion of zooplankton food and of spent mucus—they and the coral manage a continuous generation and transport of mucus toward the animal surface, followed by transport of the spent mucus through the mouth and into the gut, where it is reprocessed.

The microbiome in the calicodermis plays a major role in adjusting the local chemistry to optimize pH, oxygen concentration, and so on, making calcification possible. And the microbiome typical of the polyp's body wall includes bacteria that play roles in the cycling of carbon, nitrogen, sulfur, and phosphorus to facilitate the metabolic processes of the coral animal and of its algal symbionts. In other words, just as the coral reef neighborhood is a place occupied by numerous species of fish, crustaceans, mollusks, and echinoderms that cooperate in complex ways, using behavioral signals to communicate with one another, so the single coral colony is a much smaller set of several neighborhoods, occupied by an equally diverse collection of microbes that cooperate with one another in complex ways, using chemical signals to communicate and synchronize their actions. The larger neighborhood achieves an effective, efficient management of resources at a macro scale, while the smaller neighborhood does much the same thing at a micro scale.[23]

♦♦♦

The major tasks of corals are to grow and to reproduce. Growth requires energy, and the coral gets nearly all its energy in nutrient bars provided by its algal symbionts—the sugars, amino acids, and so on built through photosynthesis. Growth also involves laying down new calcium carbonate skeleton.

Corals do not grow *in order to* create coral reefs; they grow for the same reasons all organisms grow, but coral reefs emerge as a consequence. Without continual coral growth (plus the growth of other calcifying organisms) reefs would simply not exist, and reefs on which coral growth slows or stops are reefs that degenerate and eventually fade away. I discussed this awesome dynamic of creation and destruction in chapter 2. It's time now for the nitty-gritty details, because if the history of reef-building is an epic of awesome scale, the nanoscale details of how it gets done have a mesmerizing intricacy, precision, and complexity rivaled scarcely anywhere else on Earth—one more aspect of wondrousness.

Calcification in the ocean should be relatively simple. Calcium carbonate is weakly soluble, dissociating into calcium ions and carbonate ions, and warm surface waters are usually supersaturated in calcium ions. Other things being equal, that should drive the formation and precipitation of solid calcium carbonate (or limestone). Unfortunately, carbonate ions also get caught up in the interaction of water with carbon dioxide dissolved within it. Carbon dioxide in surface waters is at equilibrium with its presence in the atmosphere, but dissolved carbon dioxide reacts with water in a two-step dance that uses carbonate ions to produce bicarbonate ions. In the first step, a molecule of carbon dioxide and a molecule of water combine as a bicarbonate ion, releasing one proton in the process. That proton is rapidly grabbed onto by a second carbonate ion, yielding a second bicarbonate ion.[24]

The result is that carbon dioxide concentration in the atmosphere sets pH of surface waters and indirectly determines the equilibrium between solid calcium carbonate and the dissolved ions because of the way it consumes carbonate.[25] A coral, seeking to build a skeleton, confronts the problem that there is not a lot of carbonate around. How do corals deal with this? Part of the answer is that they don't calcify in the waters around them, they do it deep within, on the boundary between the calicodermis and the existing skeleton.

We now know that calcification in corals occurs because the coral and its algal symbionts (plus many bacterial members of the microbiome) work cooperatively. First, the coral microbiome aids photosynthesis by actively transporting bicarbonate from the surrounding water to the gastrodermis and its symbiocyte cells and by using a proton pump within the cell membranes of the symbiocytes to acidify the local environment of the algae living within. Measurements in one species of coral have shown the pH within symbiocytes to be ~4.0, compared to 7.4 in the epidermis and gastrodermis (ocean pH is about 8.1). In the more acid symbiocyte environment, dissolved carbon exists more commonly as molecular carbon dioxide than as bicarbonate or carbonate ions. Rubisco, the enzyme that enables the first step in photosynthesis, requires molecular carbon dioxide as its substrate. This greater prevalence of carbon as carbon dioxide within symbiocytes helps explain why photosynthesis by algal symbionts is so rapid.

We also know that calcification is intimately dependent on photosynthesis. First, the process of calcification proceeds two to four times more slowly in winter than in summer when light intensity, day length, and temperature are all greater. Second, it is also much more rapid in daylight than at night.[26] Third, calcification is dead slow in corals that have lost their algae through bleaching and in juvenile corals that have yet to gain symbionts.[27] The algal symbiont and its photosynthetic activity are crucial to calcification by the host coral.

Although algae play one obvious role in calcification simply by providing those energy bars the coral will need to build its calcium carbonate, a second role is to increase the availability of carbon beyond that dissolved in the surrounding waters—metabolism of those energy bars by the coral releases carbon dioxide within the tissues of the coral. This can be used in further photosynthesis or transported to sites of calcification.

Corals, helped by various bacteria within the microbiome, also transport bicarbonate ions into the tissues of the calicodermis and into the boundary between that and the skeleton, pushing pH in the boundary higher than that in the ocean water outside. At the boundary, bicarbonate is converted to carbonate, facilitating calcification.[28] Typical daytime rates of calcification are 0.1 to 0.2 milligrams of calcium carbonate per square centimeter of coral per hour.[29]

Putting this all together, even though photosynthesis is speeded up at lower pH while calcification is speeded up at higher pH, the coral, its algal symbionts, and many of its bacterial symbionts work together to pump protons from one place to another, to transport bicarbonate ions across cell membranes, and to facilitate adjustments in the concentrations of carbon dioxide and oxygen as needed at sites where photosynthesis or calcification is taking place. In these ways, these superficially competing, or antagonistic, chemical reactions are both facilitated. The result is an extraordinary collaboration among symbiotic species. Together, they achieve exceptionally high levels of photosynthesis in a desert and rates of calcification sufficient to build majestic reefs thousands of meters thick, hundreds of meters wide, and thousands of kilometers long—the largest creations by any organism other than humans on this planet. There is nothing trivial about coral chemistry.

♦♦♦

Coral reefs are remarkable on many fronts. Maybe I have gone too far into the science, but it is through science, and the excitement of discovery, that I have learned to love coral reefs. I can only discuss them from my own perspective.[30] There are lots of other stories I could have told, and plenty of stories that others could tell far better than me, so these eight chapters are a tiny dip into what reefs really are. They are amazing ecosystems, and our world is richer because of them. It's time now to ask that vexing question: So what good are coral reefs anyway?

NINE

What Good Is a Coral Reef?

I don't actually remember the first time I heard someone say, "What good is a mosquito anyhow?" I do remember my immediate and instinctive response—"Who in hell do you think you are to ask such a question?" I did not verbalize that response, but I thought it. Such responses guarantee that I'd make a lousy property developer. Give me 10 hectares of land and funding to build houses, and I'd begin by looking at the land and then trying to figure out where to place a couple of houses without disturbing anything too much. I might even conclude that this particular 10 hectares should not be used for housing. I certainly would not start by visualizing the 10 hectares as a parcel of land that I was free to scrape and mold into any shape I pleased so that I could arrange the maximum number of houses, roads, and so on across its surface. If someone unwisely hired me as a property developer, I would do okay with property in the form of an old, relatively flat, or gently rolling field. Anything else and I'd begin by looking for ways to retain as much as possible of what was already there. That clump of trees, those rocky cliffs by the stream—and I'd end up with far too few houses.

When I was eleven, a friend and I happily spent several days working hard with an ax to fell a number of aspens, on property we did not own, turning them into logs to build a log cabin; we abandoned the project when the walls were a modest 1 meter high. I think that episode was followed by the one involving slingshots and songbirds—another enterprise that ended abruptly once I had perfected my technique sufficiently to kill a couple of sparrows. The point is, I am not a total wimp when it comes to modifying the natural world. But I do begin with the assumption, which came to me from I know not where, that the living veneer that clothes our planet mostly deserves some respect.

I don't think I've ever been asked, "How much is a coral reef worth?" But if I had, I would have had the same reaction as I do to questions about mosquitos. Some questions don't really deserve answers, and I hope that by this point in our journey I've convinced you that coral reefs deserve our respect and appreciation, if not also our love.

Yet to most people, questions about the economic value of nature are important. It's a way of relating to nature that I do not understand, but I acknowledge that it's an approach in widespread use. Although people may have asked this question a hundred years ago or more, it seems to have grown in importance as our detrimental impacts on coral reefs have grown. The conservation, or sustainable management, of coral reefs depends on the reefs having perceived economic value. At least, that is what the great majority of people, including many conservation scientists and ecological economists, believe.

Although many of us recognize value in coral reefs or other parts of the natural world without requiring that value to be specified in economic terms, talking in terms of economic value can raise funds to support conservation and can justify that conservation to many who would not otherwise be supportive. As we steadily degrade the planet, the challenge of conservation grows, and the need to assign economic value seems also to grow.

Value, whether economic or otherwise, is a human construct. Without people to set a value, nothing on this planet is of more intrinsic worth than anything else. Never mind the nonliving aspects of landscape—that valuable hill or this even more valuable mountain. When it comes to living things, evolution proceeds unimpeded by concepts of value; its results arise because of simple principles and causal processes that have nothing to do with relative value and everything to do with time, chance, and fitness for whatever the present environment brings.[1] There is no top to the tree of life, just a multitude of tiny terminal branches, and some of those branches will bear new buds in due course. Those fitter branches may be closely or distantly related to each other, and what is fittest today need not be fittest tomorrow or next year or century, or in the next valley over. Bringing it back to coral reefs, just because coral reefs of one type or another have been around since the mid-Ordovician, 470 million years ago, does not make them any more or less valuable than any other ecosystem to ever have existed on the planet. Coral reefs just are.

♦♦♦

Yet it is right to talk about the economic value of a coral reef. A value in dollars can be informative for those less attached emotionally to reefs, and reefs do have real economic value for humanity. The challenge is to quantify that value and build it into economic decision-making—unquantified statements about value don't easily fit on cost-benefit spreadsheets used by hard-nosed economists to make decisions. And deciding what actions a society will take, or what other actions it will cease taking, usually requires hard-nosed economists somewhere or other for the decisions to get made.

Our global economy has developed without taking much account of environmental costs or benefits deriving from our activities. For most of our history, it did not really matter if most of us treated the environment as a larder full of things for us to use for free and a refuse container for those things we wanted to throw away. But as we have become more numerous, and as our economy has grown, both in absolute and in per capita size, our impacts on the environment can no longer be ignored. To incorporate environmental effects into economic decision-making, we need to bring the environment onto the balance sheet used in making those decisions. To do this requires that we start with value.

Many people have undertaken to explore ecosystem value and to specify it in financial terms. Doing this is a primary objective for the growing research field of ecological economics. It grows out of a belief that if the value of an ecosystem can be specified in monetary terms, that will encourage including, in economic decision-making, the benefits of sustaining, or the costs of degrading, that ecosystem. I'm not sure this argument sways many people not already sympathetic to ideas of conservation or sustainability. In fact, I fear that there is a wide gulf between the worldviews of people who see implicit value in sustaining the natural world and those of people who think primarily about the world of commerce, profit and loss, depreciation on assets, and wealth management. I'll tackle this issue in the final chapter; for now, let's look at ecosystem value and ways of measuring that.

A coral reef is one type of ecosystem occupying a small portion of coastal ocean. As such, it is a store of natural capital able to provide a flow of goods and services of certain kinds to people. Those goods and services have value for people; they are the production from that natural capital, which also has a value. The natural capital value may grow, remain constant, or decline over time. The goods and services are a dividend or profit with a value per year. Again, that value may be constant year after year, may grow, or may decline as the reef becomes degraded.

Important here is the notion of sustainability; it is widely, if sometimes grudgingly, acknowledged that we should manage our uses of ecosystems so that natural capital is preserved for the future. The rate of extraction should not be more than the rate at which resources can be replenished if one plans to continue receiving goods and services into the future. In economic terms, the rate of extraction should not deplete the natural capital. The value of that natural capital usually is not included when reporting on ecosystem value; sustainability is assumed, and the (current) value is that of the flow of goods and services. The United Nations framed this notion of sustainability in human terms in the 1987 Brundtland Report, and subsequently in the 1992 Rio Declaration—sustainable use ensures continued availability of those goods and services for people in the future and is therefore an appropriate goal for humanity.[2] Again, without humans to provide value there is no reason why we should not ravage the world without regard for the future. Nature, the environment, the biosphere, the planet—none of them care a whit whether we act sustainably or not.

The simplest way to value a reef adds up the current economic value being derived from use of identifiable, extractable resources per hectare—the fish, coral jewelry, curios, and so on. A little reflection quickly tells us that this resource-based value is only a small portion of the total value, because reefs also provide nonextractive ecosystem services such as coastal protection with value that is much harder to estimate. This is generally true of all ecosystems, and many different ways of categorizing and quantifying the services provided by ecosystems have been developed. When the Millennium Ecosystem Assessment published its report *Ecosystems and Human Well-Being* in 2005, it categorized all ecosystem services into Provisioning, Regulating, Cultural, and Supporting services, and this categorization has been widely adopted.[3]

Provisioning services are the flow of products we obtain from an ecosystem, such as fish, grain, and other wild and cultivated foods, wood and fiber, pharmaceuticals and other organic chemicals, freshwater, and biofuels. Our economy also uses a range of other natural products including metals, other inorganic chemicals, and fossil and nuclear fuels. While there can be important ecological impacts of extracting or using these products, they are not produced by the ecosystem, at least over timescales relevant to economic decision-making or the lives of people or nations.[4] Such products are not usually included in the provisioning services of an ecosystem. Provisioning services are the flow of ecosystem products through markets, or potential markets, and their value can usually be estimated by reference to prices paid in those markets.

Regulating services are the suite of processes occurring in an ecosystem that provide human benefit by regulating environmental conditions within and beyond that ecosystem. These are often specific to particular ecosystems. Salt marshes and mangrove systems trap, filter, and process pollutants from the land that would otherwise contaminate water bodies beyond them. Forests and marshes sequester carbon that would otherwise leak to the atmosphere and accelerate warming of the climate. Many ecosystems have important roles in the purification of water. Most support diverse assemblages of insects and other species that together provide a pollination service vital to agriculture. Some ecosystems, such as coral reefs, provide important protective services to people and infrastructure nearby. One approach to valuing regulating services, in cases where an engineering solution exists, is to determine the cost of doing the regulation through engineering. Such cost estimates are often very high.

Cultural services are the nonmaterial benefits people obtain from ecosystems through spiritual enrichment, cognitive development, reflection, recreation, and aesthetic experiences. Included here is the value derived from recreation and tourism, and also the value an individual human derives from just knowing, for example, that a coral reef exists. Cultural services include a mix of services for which markets exist (tourism) and ones without a market (value from knowing that coral reefs exist).

Supporting services differ from other ecosystem services in that they affect people only indirectly or over long periods of time. They include soil formation, photosynthesis and primary production, and the cycling of water and nutrients. These are all processes on which ecosystems depend in producing their provisioning and regulating services. They are also particularly difficult services to assign value to, yet like provisioning, regulating, and cultural services, supporting services exist in the context of value for humans even if the links to humanity are long and well hidden. We value ecosystems only for the services they provide to us.

Over the past twenty years, ecological economists have debated both the relative merit of different approaches to valuation and the benefit of assigning monetary value to the range of goods and services derived from ecosystems.[5] Some fear the commodification of nature that assigning monetary value would bring. That danger may be real because conventional economics generally assumes that all sources of value are substitutable. A forest, for example, can be viewed as a store of value in timber. Harvesting it all and investing the proceeds in Apple or Amazon might yield far greater monetary value in future years than if the forest was sustainably harvested. That may be economically prudent, but is it environmentally wise? Is a tree replaceable by shares in a tech stock?

Other economists argue that if realistic monetary values are not provided for ecosystem goods and services, our economy will never succeed in properly accounting for the cost of environmental degradation or the benefit of sustainable environmental management. One problem I see is that the notional monetary values of the flows of goods and services from many natural ecosystems are so large that they fail to convince those not already used to valuing nature or the services it provides. They are too big to be believed!

Regulating, cultural, and supporting services mostly do not operate through markets, and assigning quantitative value to them can be problematic, but progress has been made. Efforts to value the goods and services derived from ecosystems began well before the Millennium Ecosystem Assessment, and effort accelerated following the first attempt to provide a global total value in 1997. In that year, Robert Costanza of the University of Maryland and a dozen colleagues from the United States, the Netherlands, and Argentina reported in *Nature* that the total global value of ecological goods and services in 1994 was about $33 trillion per year, a number almost twice the global gross national product (GNP) of $18 trillion per year.[6] Subsequently, Costanza, Rudolf de Groot of Wageningen University, Netherlands, and others worked as part of the TEEB (the Economics of Ecosystems and Biodiversity) effort to build a global database of values of specified ecosystem services for each of ten biomes.[7] In 2012, de Groot and sixteen globally distributed colleagues, including Costanza, reported in the journal *Ecosystem Services* that the total annual value of ecosystem goods and services ranged from a low of $491 per hectare in open ocean ecosystems to $351,000 per hectare for coral reef ecosystems (2007 dollars). In 2014, in *Global Environmental Change*, Costanza, de Groot, and colleagues reported the change in global value of ecosystem services between 1994 and 2011. Their calculations revealed a nearly threefold increase in the overall value of ecosystem services from $45.9 trillion per year (value converted from 1994 to 2007 dollars) in 1994 to $124.8 trillion per year in 2011 (also in 2007 dollars). For comparison, global gross domestic product (GDP) in 2011 was $75.2 trillion.[8]

That threefold increase in value of ecosystem goods and services was due primarily to better (more comprehensive) accounting rather than real growth. However, it masked an unanticipated loss of $20.2 trillion per year due to reductions in the areal extent of more valuable ecosystems over the fourteen years! Across the globe, substantial losses occurred in extent of tropical forests, wetlands, tundra, and coral reef, and substantial gains in (less valuable) deserts and subtidal algal beds. Those changes are largely attributable to human activity, and the net loss of value shows that, at this global scale, the ideal of sustainability was not being achieved. That the world was largely unaware of the economic cost of loss of rain forests, reefs, and so on, shows how little a part the concept of ecosystem value plays in our thinking. Twenty trillion dollars per year gone up in smoke over fourteen years, and we didn't even notice.

♦♦♦

The TEEB database compiled by de Groot and colleagues includes ninety-four estimates of total ecosystem service value for coral reef systems around the world. They range from just under $37,000 to over $2 million per hectare per year, due to paying attention to different sets of services, varying methods for valuation, and real differences in value among reefs.

Studies undertaken since the work of Robert Costanza and Rudolf de Groot further define the range of economic value for coral reefs. The Bermuda Department of Conservation Services published a detailed evaluation of Bermudian reefs in 2010.[9] It reported a total economic value of the flow of reef goods and services to be about US$722 million per year (2007 dollars), or 12 percent of the country's GDP. This was composed primarily of $406 million for tourism and $266 million for coastal protection. The valuation methods used do not translate easily into value per hectare of reef, but the study considered the approximately 400-kilometer-square area of platform less than 20 meters deep, including substantial areas of sand. If half the platform is composed of reef habitat (a plausible guesstimate), the total economic value approximates $36,100 per hectare per year, at the low end of the reef values reported by de Groot.

In 2017, Deloitte Access Economics published an updated economic valuation of the Great Barrier Reef.[10] This evaluation included A$6.4 billion per year as its contribution to Australia's GDP, 64,000 full-time-equivalent jobs, and a total asset value of A$56 billion. The A$6.4 billion annual value is the latest in a series of economic valuations done by Access Economics in recent years. There has been a slow growth in this value from A$4.5 billion in 2004, due almost entirely to growth in tourism.[11] The estimate is again not easily converted to a value per hectare of reef; methods used differ from those used by de Groot, Costanza, and others in several ways.

The estimate focuses on market values to estimate value of the current flow of goods and marketed services to yield the A$6.4 billion estimate and the number of jobs. The estimate includes the total asset value, a measure of the natural capital value represented by the Great Barrier Reef, calculated assuming a thirty-three-year lifetime and depreciation at 3.7 percent per year, and based on perceived value of the reef by Australians. (They use a variety of willingness-to-pay approaches to capture perceived value, over and above money actually spent by Australians visiting the reef, to produce this A$56 billion estimate and note that it does not include aesthetic or spiritual value or value ascribed by [Aboriginal] traditional owners.) One can argue both with how perceived value was captured and with the lifetime and depreciation rates employed—this is economics, not physics—but it yields a defendable number.[12] Finally, the valuation includes a discussion of the (presumed considerable) aesthetic, cultural, and spiritual value, particularly as espoused by traditional owners, and a separate discussion of the (also presumably considerable) value of the reef as a contribution to Australia's brand. Although neither of these types of value are quantified, the report argues for their inclusion when considering overall value, and I'll return to this point later.

This is only a glimpse of the range of valuation studies available. Some look at total economic value per year, others report value of specific services such as tourism or coastal protection, and some such as the Deloitte Access Economics evaluation of the Great Barrier Reef also include natural capital value. Methods vary greatly, and estimates of value also vary, both because reefs differ in value and because of variations among methods. It is undeniable, however, that coral reefs, particularly those that protect nearby coastal communities, and those that are targeted by tourism, can provide substantial economic returns to countries that possess them. A focus on these more easily quantifiable goods and services leads to impressively large valuations. We could attempt to go further. We might recognize their existence value as by far the richest stores of biodiversity on the planet or the supporting services in their efficient nitrogen fixation and nutrient cycling, which create productive oases in the otherwise nutrient-poor tropical ocean. Or we might include the cultural, spiritual, or aesthetic benefits they have for many people. As we do so, valuations will become even larger, but also fuzzier. For present purposes, let's just accept that many coral reefs have been valued in the tens to hundreds of thousands of dollars per hectare per year, while remembering that these values are underestimates because not all perceived forms of value are included. Despite reefs being incredibly valuable to humans in many ways, most of that value has yet to find its way routinely into statements of monetary value.

So, reefs can be valued in monetary terms, and the annual value derived can be quite large. What now? Valuation can become an argument in favor of investing government and/or private funds in their sustainable management. This is the approach taken in the Deloitte Access Economics report, *At What Price?* It is written in a way that clearly advocates for action to conserve the Great Barrier Reef. The final paragraph of the report closes with, "But more than just getting the policies right and investing wisely, understanding the total value of the Reef shows us what is at stake. And when called on, it is this knowledge that allows us to make it clear that the Great Barrier Reef's protection is not only an Australian priority, or an international one—it is a human one."[13] This theme is common to valuation studies elsewhere.

A more limited and quite different approach was used in a 2011 evaluation of tourism value in Jamaica.[14] Ben Kushner and colleagues related reefs to sand production and sand production to the maintenance of white sandy beaches. They then determined the importance of beaches to tourists. Knowing the annual economic value of tourism, they were able to assess the value of coral reefs in sustaining beaches in Jamaica, but they did not report this value directly. Instead, they used information on current rates of beach erosion and models that projected increases in erosion as reefs degraded. They were then able to report the tourism value that would be lost if reefs continued to degrade, and therefore the value of acting to prevent reef degradation. Such an argument could be used by policy makers to increase the priority of reef conservation in government circles.

♦♦♦

The morning of December 13, 2003, was sunny and hot, like all days in Dubai. I was standing at the Nakheel site office on the coast, looking out to where Palm Jebel Ali was under construction. Palm Jebel Ali, a giant, 7-kilometer-diameter crescent island surrounding a complex, vaguely palm-like trunk extending out from shore, was being built by dredging great quantities of sand from the immediate vicinity and piling it up to build islands where none had existed before. I could see most of the crescent across the water as low sand dunes stretching across the far horizon, with giant cutter dredges working here and there shooting enormous plumes of water and sand like rooster tails high in the air and over to the growing island. Parts of the trunk were complete, but not yet extending to connect onshore. I knew that underneath the inshore, southwestern portion of this construction site had once existed the Jebel Ali Reef, the only significant inshore coral reef on the Dubai coast. That now buried reef had been protected since 1998 by the Jebel Ali Marine Reserve.

So much for permanent protection. I was in Dubai, on the first of many visits over the next seven years, leading a project funded by Nakheel, the giant quasigovernmental property developer. I was there to help Nakheel learn to manage the waters surrounding its enormous, offshore island complexes, of which Palm Jebel Ali was only one. In this project I came face to face with people, and a corporation, with little concept of the value of natural environments or of the idea of conservation. During that same first visit, my colleague, Jake, and I were standing on the then similarly sand-dune-like trunk of nearby Palm Jumeirah. As I remember, Jake put a simple question to our guide, Imad, then head of Nakheel's Environment Department. Something like: "Have you thought about conservation for some of the native ecosystems?" Imad began with how very committed to conservation Nakheel was. Then, waving his hand toward some shrubs in a small garden plot, he told us that those plants had been brought from many countries around the world and were planted there to determine which will be best for landscaping on the finished islands.[15] That reply told me how much work we had in front of us!

The attitude to nature reflected in Nakheel's property development activities is not unique to Dubai or to the developing world more generally. Many of us scarcely ever think about the value of natural ecosystems or the value gained by sustaining or conserving them. That reef, casually buried with no serious discussion of whether this mattered, was composed of some of the most heat-tolerant corals on Earth. These were corals that had found ways to survive water temperatures warmer than most marine regions ever experience and salinities approaching 40 parts per thousand (ppt); most sea water has a salinity of 34 to 36 ppt. Corals in Dubai had bleached in past years, but the survivors might carry gene combinations that would be particularly valuable as the planet warms. They were buried to build sand islands.

Burying a reef is extreme, but governments regularly neglect their reefs. Although the numerous valuation studies, and the values derived, are always impressive, these do not appear to have influenced coral reef policy by governments as much as one might have expected. This is surprising because most valuations focus on tourism, fisheries, coastal protection, and other services that can be evaluated relatively easily in monetary terms. The monetary values are real and could be expected to sway attitudes of policy makers. Yet despite coral reefs being, per hectare, the most valuable aquatic ecosystem out there, governments still tend to drag their feet when it comes to investing appropriately in their management.

An embarrassingly clear example of this issue is the Great Barrier Reef. To its credit, Australia forty years ago created the necessary underlying legislation and established the Great Barrier Reef Marine Park. That was a complex political process because Australia is a federal democracy.[16] Australia has also committed significant funding for management operations within the Great Barrier Reef Marine Park. The 2015–2016 operating budget for its manager, GBRMPA, was A$28 million.[17] Nevertheless, in recent years, Australia has been resistant to curtailing either the expansion of coal mining or the expanded export of bulk coal through coastal ports within the Great Barrier Reef Marine Park. Australia has also been reluctant to undertake real steps to reduce its greenhouse gas emissions in line with the 2015 Paris Agreement. Only strong political pressure from Australia prevented UNESCO from listing the Great Barrier Reef World Heritage as "world heritage in danger" in mid-2017, and UNESCO continues to express concern about Australia's long-term management plan for the reef (which includes no action to mitigate climate change). Clearly, the A$6.4 billion in annual economic benefit has not yet swayed politicians very far.

I alluded earlier to the general failure of ecosystem valuations to galvanize conservation action, and I suspect it has to do with the difficulty people face in adjusting their individual mind-sets to accommodate attaching a large monetary value to something (the environment) that has always been there and has never had to be valued before. There is a tendency to disbelieve the numbers. A second, certainly important, factor is that the value ascribed to a reef (or to any ecosystem) is a value, only part of which is delivered to specific sectors of the community—the tourism operators, the fishery. The rest of the value is distributed broadly to the community but largely invisibly—the coastal protection or the flow-on economic benefits from a vibrant tourism industry. If an individual or an entire community sector (such as the coastal farming community) not receiving direct economic benefits from a reef happens to be doing something that damages the reef, that individual or community sector would have some costs in stopping doing that damage and would gain no direct economic benefit from doing so. This is not an issue limited to reefs; it's one of the major reasons why we so frequently manage environments unsustainably. Repairing or preventing degradation requires that individuals doing the damage incur costs to rectify their activities, while benefits to so doing are (invisibly) shared across society.

There is one more issue to reflect on here, and one I struggle with. Reefs being used by people are more valuable than remote, seldom visited ones. Ecological economics necessarily must deal with value as seen in human economies, and reefs that are not used do not have any of this value. In a sense, the remote, pristine coral reef (a few may still exist) is a lot like that tree that fell down in the forest when nobody was around to hear it. Did that falling tree make any sound?

Somehow, I suspect that humanity will have to find a way of valuing the unused as well as the used if we are truly going to relate effectively to the natural world. We also need a way of fairly quantifying nonmarket value as well as market value.[18] In this respect, the Deloitte Access Economics valuation, even though it attached no monetary value, may be setting the stage for a more inclusive way of valuing reefs and other natural systems by raising the issue of the aesthetic, spiritual, and even iconic value of the Great Barrier Reef. Reefs should be considered valuable just because they are. Still, I am getting ahead of myself. It's time now to accept that reefs can be very valuable to us and to turn to the question, What are we doing to coral reefs?

TEN

Why Don't We Seem to Care about Coral Reefs?

In February 1968, Elmer Robinson and S. C. Robbins of the Stanford Research Institute presented their final report to the American Petroleum Institute. In it they identified carbon dioxide released through the use of fossil fuels as a major atmospheric pollutant. They warned of potentially dangerous environmental effects and predicted that concentrations in the atmosphere, then at 323 parts per million (ppm), might reach 400 ppm by 2000.[1]

In June 1968, I received my Ph.D. at an outdoor convocation redolent with the scent of flower leis in Honolulu and began preparing to move to the University of Sydney to begin my studies of the Great Barrier Reef. I was completely unaware of the concentration of carbon dioxide in the atmosphere, that it was becoming more abundant, and that it would materially affect coral reefs and coral reef science before my career came to its end. I did not even know that the first instruments deployed to provide a continuous measure of carbon dioxide concentration had been high atop Mauna Loa since 1958, scarcely 300 kilometers from the University of Hawai'i campus. Nor were my coral reef colleagues aware; we all still thought coral reefs were really neat systems to study and assumed they would always be here for us. The American Petroleum Institute staffers likely did not think much about coral reefs.

The report by Robinson and Robbins was an early sign that the fossil fuel industry was aware of its potential to reshape climate, with global environmental consequences. I mention it because the question of what is happening to coral reefs is increasingly a story about climate change due primarily to our burning of fossil fuels. The fossil fuel industry, with Exxon-Mobil in the lead, has conducted a highly effective campaign to deny the evidence that we are warming the planet. Exposed in glaring detail by two reports in 2015, that effort began in the late 1960s and continues today. That program of denialism has significantly affected the pace at which climate policies have been enacted.[2]

In 1982, as the world entered the strongest El Niño ever recorded until then and the surface waters of the Eastern Pacific became very warm, the manager of Exxon-Mobil's Climate Affairs research program sent a confidential memo to senior management titled "CO_2 Greenhouse Effect."[3] The memo reported that carbon dioxide concentration in the atmosphere was rising, projected a likely rise in global average temperature of about 3°C toward the end of the twenty-first century, and stated that mitigation of this greenhouse effect "would require major reductions in fossil fuel combustion." (The memo also suggested that reducing fossil fuel use would be premature until there was more certainty of the climate trend and impacts.) In 1983, as the El Niño ended, Exxon reduced its funding for climatological research from $900,000 per year to $150,000 (out of a total research budget of $600 million) and ramped up efforts to sow doubt about the science of climate change. The 1982–1983 El Niño resulted in the first geographically extensive mass bleaching of coral reefs ever seen.

In 1982–1983, I was studying fishes on the Great Barrier Reef, and if I thought of coral bleaching at all, I thought of it as an interesting problem for corals. In May 1983, carbon dioxide concentration in the atmosphere atop Mauna Loa reached 346 ppm.

In 1997–1998, the world went through an El Niño that was as intense as the 1982–1983 version but slightly longer in duration and until 2016–2017 the strongest El Niño on record. It led to a series of mass bleachings of coral reefs that extended around the world—the first such circumtropical occurrence of this phenomenon. At the same time, Exxon's chief executive officer claimed (in an address to the Fifteenth World Petroleum Congress in Beijing) that global climate was not changing, and even if it was, fossil fuels were not to blame. Atmospheric carbon dioxide concentration reached 369 ppm in May 1998. At a conference at Nova University, Fort Lauderdale, in 1999, I finally realized what had been happening to corals around the world and began to understand what this meant for reefs, for our planet, and for us. As I write, carbon dioxide concentration has topped 417 ppm (during May 2020).

♦♦♦

In 1982–1983, the El Niño and resultant warming led to the widespread bleaching of corals and death of 70 percent to 90 percent of all corals on reefs of Pacific Panama. Bleaching and death were also extensive in the Galápagos and from Costa Rica to Ecuador along the Pacific coast of Central and South America. This was the first mass bleaching event ever recorded. Numerous similar events have now occurred, and circumtropical bleaching events, involving extensive bleaching of corals on reefs around the world, have occurred four times, first in 1997–1998 and most recently in 2014–2017. Since the early 1970s, the accumulation of destructive events caused by local or more wide-ranging human impacts, including climate change, has overwhelmed the capacities of reef corals to grow and multiply. Caribbean reefs and the Great Barrier Reef have each seen more than a 50 percent reduction in the coverage of living coral present.[4]

The story of what we are doing to coral reefs can be long and depressing. I covered it in *Our Dying Planet*, when it was a few years shorter but still depressing to me. The ecosystem to which I have devoted my career is very likely to disappear from the planet about the same time as I do. Dust to dust. And yet, if I've been successful so far in helping you to appreciate coral reefs, now is the time for you to understand what we are doing and to join the battle to get us moving to rectify the damage.

The coral reef story is inextricably linked to our unsustainable use of the planet and increasingly to our impacts on climate. I'm only going to skim here, because I want to focus on why we are ignoring this unfolding calamity. At present, humanity consumes natural resources at about one and a half times the rate at which our planet can produce them. The waste products of our economies and our individual lives place nearly impossible burdens on natural systems, polluting water, soil, and atmosphere. Many of these changes, such as those in climate, are now far more rapid than at any time since the earliest dawn of agriculture, and our impacts will likely increase substantially as our population grows from today's 7.6 billion to about 10 billion by 2050 and as our average standards of living increase across the world. It's not just about coral reefs.

Our growing impacts have severe consequences for the biosphere, for many individual species and ecosystems, and for our own lives. While the ideal of environmentally sustainable human development has long been aspired to, the global trend has been one of continuing degradation of environmental quality.[5] We now face challenges of existential proportion, because we've been changing the world in bad ways—ways that make it a less wonderful environment for creatures like us. We are radically altering the pattern of habitats across the planet while taking ever more land (now 75 percent) for agriculture, industry, and human settlement. We are polluting the air, water, and soils, often with chemicals not known before we invented and manufactured them. We are overexploiting freshwater resources and those wild stocks of fish, forests, and wildlife we value. And we have been causing the extinction of numerous species. These direct impacts of our actions have numerous knock-on, indirect impacts such as the changing climate, rapidly rising sea level, loss of polar and alpine ice, growing deserts, an acidifying and deoxygenating ocean, and rapidly degrading coral reefs.[6]

Human-caused extinctions have been a feature since the late Pleistocene, but our pace in finishing off other species has increased substantially. We may have triggered the sixth great mass extinction on the planet—average extinction rates for vertebrates are now estimated to be between one hundred and one thousand times higher than typical over the grand sweep of geological history. The last great mass extinction took place at the end of the Cretaceous age, 66 million years ago, and took out the last of the dinosaurs, among many other creatures. We have no idea what will disappear in the sixth extinction, although most nondomesticated vertebrates are high on the list. Only hubris tells us this does not matter.

In the oceans, the extent of human impacts began later but is now catching up to that on land. No part of the oceans remains unfished, and we have reduced the standing stock of fishery species by 90 percent over the past hundred years. Bottom trawling, which resuspends 22 metric gigatons of sediment per year, has substantially degraded benthic habitat over 20 million square kilometers (75 percent) of continental shelves, significantly reducing the productive capacity of these environments. Chronic pollution has generated more than five hundred dead zones in coastal waters, and the Great Pacific Garbage Patch, three times the size of France, contains about 80,000 metric tons of floating plastic debris. Atmospheric carbon dioxide dissolving in the ocean is reducing pH ten times faster than at any time in the past 66 million years, threatening the existence of many marine species.[7] As well as bleaching reefs, our warming of the planet has already ensured that sea level will rise 1–2 meters by 2100 and continue to rise for at least the next thousand years.

Coral reefs are dependent for their existence on exquisitely sensitive creatures. Through their calcifying activities, corals provide the intricate structure that constitutes the reef and generates the habitat for thousands of other species of life. Corals have narrow tolerances for salinity, turbidity, light intensity, pH, nutrient concentrations, and temperature. Humans have always damaged reefs, overfishing them, fishing them using methods that damage the reef structure, polluting them with runoff from coastal locations, smothering them in sediments caused by inappropriate coastal construction, mining them for building materials, trampling over them at low tide, or turning them into runways connecting isolated atolls to the global air transport network. Until recently, human damage was locally caused and could be locally corrected.[8] However, our emissions of carbon dioxide and other greenhouse gases are now killing reefs globally by changing ocean temperature and pH.

Reefs have no chance to recover when bleaching happens every year or so, yet the Great Barrier Reef bleached severely in early 2020, the third serious bleaching there in five years. Given the pace at which temperature is increasing, the prognosis is dire—many of us see little chance that reefs resembling those of the 1960s will persist past 2050. If this future comes to pass, we will have witnessed the elimination of an entire ecosystem and the regional or global extinction of many thousands of species within a single human lifetime.[9]

There is now a real risk that human activities could push the Earth system outside that state in which it has existed since the retreat of the last Pleistocene glaciers, destabilizing it in the process. By destabilization, I mean that we could unwittingly tip the planet toward a new equilibrium different to the one it has existed in through at least the past eleven thousand years. One way in which we may well have already done this is by warming the climate sufficiently that the melting of glacial and sea ice has caused enough of a decrease in Arctic ocean albedo (reflectance), and/or sufficient releases of permafrost methane, to set up positive feedback for further melting. If so, no matter what we do to rein in our own carbon dioxide emissions, melting will continue until all permanent ice is gone, driven by mechanisms over which we have no control, and raising sea level 75 or 80 meters. A slow melt of glaciers over the next several thousand years is gradual enough that we could adapt to it, but it would be costly, societally disruptive, and would ultimately lead to a very different planet. Other shifts of equilibria could be a lot more abrupt. In any event, a Holocene planet is the only planet civilized humans have known, and the precautionary principle suggests that we'd be wise to curb our environmentally destructive behavior and learn to live within the parameters set by the planet, as governed by the laws of physics, chemistry, and biology.[10]

Until now, we have seldom had to act deliberately to achieve common goals for planetary conditions, and there are reasons to be concerned that we may not succeed in this great challenge. We may never develop a will to change our ways; we may leave it so late that the planet's ecology will deteriorate so far that even the mighty *we* will not be able to turn it around; or we may simply underestimate the challenge facing us. Judging by our slow pace in simply coming to grips with the extent of this problem, since coral reefs first began alerting us by mass bleaching episodes in the early 1980s, it is difficult to remain optimistic about our chances of success.

♦♦♦

By now, anybody who pays any attention to environmental news knows that coral reefs are being seriously degraded (fig. 23). Yet this news has had little effect on people despite the rapid and conspicuous collapses of reef communities that follow severe bleachings. Understanding why might help

Figure 23a. The impact of bleaching on coral reefs is well represented by this pair of photos of nearby, originally quite similar sites off Mo'orea, French Polynesia. A severe bleaching event hit Mo'orea's reefs after the photo above was taken on March 4, 2019, and before the photo in fig. 23b was taken on May 17, 2019. Photo © Luiz Rocha, California Academy of Sciences.

us understand our collective amnesia regarding the destruction we are causing across the ecosystems of the planet.

The rate at which coral is disappearing is proportionally far more rapid than the rate of loss of rain forest and is clearly incompatible with the continued existence of coral reefs. Scientists are united in attributing major portions of this loss of coral to the effects of climate change-induced bleaching. The most recent (2018) report from the Intergovernmental Panel on Climate Change, an authoritative but cautiously conservative body, reports that "even achieving emission reduction goals consistent with the ambitious goal of 1.5°C under the Paris Agreement will result in the further loss of 90% of reef-building corals compared to today, with 99% of corals being lost under

Figure 23b. The difference in appearance is stark. Over subsequent months, most of the bleached colonies died, became overgrown with algal turf, and began to break down, yielding a topographically simpler reef supporting little live coral. Such has been the sequence at numerous locations around the world since the early 1980s. Photo © Luiz Rocha, California Academy of Sciences.

warming of 2°C or more above the pre-industrial period."[11] In other words, those who know the science believe reefs are virtually toast this century. One might expect people more generally to have become very concerned; and yet, they haven't.

We reef scientists have long wondered why what is happening to reefs has not mobilized deep concern across the world to act on climate change. Indeed, I vividly remember that Fort Lauderdale conference in 1999 where the effects of the strong 1997–1998 El Niño were a hot topic. Conversations kept coming back around to the hopeful expectation that the world's first circumtropical mass bleaching episode would be a strong wake-up call to the world and would jump-start the effort to rein in emissions of greenhouse gases. To us, the link was obvious, and the consequences of ignoring climate change were going to be devastating in many ways, far beyond our coral reefs. Reefs were just the canary, doomed to suffer first and thereby warn the world.[12] But it did not turn out as we expected.

Ever since, reef scientists and managers have struggled to articulate the story of coral reef decline in ways that will more effectively capture people's attention and lead to strengthened policy on climate as well as more effective reef conservation around the world. We've provided detailed case studies of bleaching events. We've explained the links between rising temperatures, bleaching, coral mortality, and reef degradation. We've used powerful models to project likely futures. We've helped document the enormous value of coral reefs, economically, aesthetically, and biologically. And we've advocated for action, locally and globally, that would help sustain coral reef systems. All seemingly to little avail.

As I have pondered, as I have watched reactions when I speak to citizens on coral reefs or climate change, and as I have lain awake nights wondering why most people just don't get it, I've identified, talked about, and then discarded one explanation after another: Most people have not yet been confronted with the evidence. Most people do not understand the science. Most people have been led astray by the powerful denialist campaigns of disinformation and contradiction. Most people have belief systems that just do not permit the possibility of existential crises. Most people understand what is happening, but we are too selfish to make any changes to our lives.

Each of these possibilities is valid, to a degree, for some of us. But even when I talk about these reasons for failing to see what is happening, my explaining does not seem to sway people! While denialist campaigns have been effective, there is something more—deep aspects of our psychology that affect how we respond to messages and incorporate them, or not, into our personal worldview. So, for one last time, perhaps just hitting my head against a wall, stick with me while I explain the ineffectiveness with which the coral reef story (and the wider climate story) has been told and the roles played by belief systems, selfish short-term thinking, and our ways of receiving and incorporating messages into personal views of the universe.

♦♦♦

Remember those tourists on the artificial island near Cancún? Few of us have ever seen a coral reef. Most of those who have, saw a reef once while on vacation, and even people living day-to-day with coral reefs just offshore have limited familiarity with what they are and how they function to sustain our lives.[13] Yet most of us would recognize a picture of a reef if we saw one and might have some ill-formed ideas about just how amazing coral reefs are. We *think* we know what reefs are, but we don't, and they are not very important to us. "Majestic," "iconic," and "charismatic" are adjectives that can be applied to reefs as well as to humpback whales, pandas, and polar bears (all of which also happen to be creatures that have been seen in their natural surroundings by rather few of us).

The coral reef story is being directed to an audience that knows less about reefs than many people assume, and the story is not being told very well. It is not having much effect. Reef scientists and managers can be blamed for part, but not all, of this failure. The media also share part of the blame. And the audience—that *everyperson* out on the street—shares the rest.

Scientists: Early in my career, one still heard the view expressed by older colleagues that academic researchers were members of an august club who communicated with each other to advance their collective understanding. Popularization, writing for the masses, talking to the media, helping make a documentary, and even writing an introductory textbook were activities best avoided if you valued your career. Fortunately, such views are now faint memories of a bygone era, but this does not mean that academia has learned how to communicate effectively with real people.

Coral reef scientists, like other scientists everywhere, seldom find the art of storytelling among the courses required during their graduate careers. Somehow, we assume that telling stories is easy and we all know how to do it. The fact is, we don't, and peer review by journals, or at conferences, seldom addresses this gap. A few of us still believe that telling stories is somehow not what a scientist should be doing; it smacks too much of entertainment. Our colleagues put up with this deficiency and force themselves to listen to our fifteen-minute conference presentations and read (or at least skim) our journal articles, even when the talks and the articles are mind-numbingly boring.[14] Real people are simply not that interested.

Randy Olson, long ago a coral reef scientist and someone who cares deeply about storytelling, recently drew my attention to the widespread ineffectiveness of our seminars, conference papers, and written communications. Properly designed, each has a structure that begins with some background (Randy calls this "and"), identifies the problem being addressed ("but"), and draws a conclusion ("therefore"). A lengthier talk or technical article will have a more complicated structure—likely an overall *and, but, therefore,* with a series of two or more subsidiary *and, but, therefore* sections within it. Each such sequence of elements builds a story arc that generates, maintains, and finally rewards interest by the listener or reader.

Although scientists are increasingly attending to storytelling in conference presentations, few of us bring this to our writing. Too often, our technical articles are a succession of *and* with no discernible story arc, just a long list of mind-numbing details. The journals we write for also go out of their way to impose a structure on articles that bears no relation at all to storytelling; in some journals "Materials and Methods" (literally "what was done") becomes "Supplementary Material" stored separately in an archive so that the main text jumps directly from introduction to results; in others, "Results and Conclusions" come ahead of "Introduction." Naturally, scientists read technical articles by glancing at the opening paragraph, scanning the figures, taking a quick look at the final paragraph, and then maybe delving more deeply. This reading is not done for enjoyment, and the articles are seldom enjoyable. Mostly, they are not memorable either.

Most scientific writing is subject to peer review, but don't expect this to improve the quality of writing; the focus is rightly on scientific accuracy and rigor. When we add deficient copy editing, a growing tolerance of slang with meaning limited to those from the same subculture as the author, and the fact that few English majors end up as scientists, it should not be surprising that technical articles are seldom models of effective storytelling and sometimes barely literate. Yet it is the stories that make a piece of science memorable. Articles that are not remembered don't get cited and might as well not have been written.

Well, okay, you say. Technical articles are intended to convey information within the science community. They were never meant to tell stories. I disagree, but perhaps more important is that we scientists also apply our story-less style to pieces of writing that are intended to reach a wider audience because we don't know any better and because we are busy. There are gloriously talented exceptions among us, but for the majority, our articles for the popular press come out as a long string of details: *and, and, and, and, and.*

Back when I began my career, the process of publication took a year or more, with manuscripts and revisions being mailed back and forth across the globe before type was finally set and a journal issue printed. I think we did a slightly better job of storytelling; I know we spent more care on each manuscript (and published far fewer). Now, in a world of instant communication, peopled by far more scientists, under far more pressure to succeed, the production of poorly written, boring articles is the norm.

Because peer-group attention is essential to maintain the stream of funding needed to do science, scientists have learned to compensate for their nonmemorable technical articles by using social media and press releases to generate buzz each time a new article appears. Many universities now have established publicity units that help with this buzz-making task; in others, the scientist has to go it alone. But how do you generate buzz? By telling effective and enticing stories! Since we scientists don't know how to do that, we adopt an easy two-step trick to create enticing copy: hype the story as new, different, the first report, a major breakthrough; and make sure the story contradicts prior studies or the current consensus.

Occasionally, there are breakthroughs in science, there are major discoveries, and there are groundbreakers that redefine our understanding. They deserve to be highlighted, shouted from the hills. But every single article coming out of a scientist's research lab? No! We all occasionally do confirmatory work or routine baseline work needed to prepare for the breakthrough. Papers reporting such work do not deserve a Hollywood treatment with searchlights in the sky and a stirring theme by John Williams. Unfortunately, in our efforts to generate buzz about everything we do, we are creating a dull background drone. We are also misleading the media and the audience.

The media: The advent of social media has been tough on the traditional media. The number of career journalists has fallen, the demands on their time are greater, and the degree to which they can afford to specialize on complex topics such as science has decreased. The pressure to publish quickly has led to verbatim line-for-line reporting of the press releases created by those hyping scientists who are not very good at telling stories. When the journalists have the time to write their own words, they latch onto the most sensational claims by the scientists and create the semblance of journalistic balance by citing one source from each side of any apparent controversy (journalists, too, are competing to be heard).

I'm not sure whether the science community deserves the greater part of the blame for the sorry state of science reporting or whether the journalists are primarily to blame for taking the bait we feed them hook, line, and sinker. Between us, we have made a mess of the reporting of science.

This mess is made worse when the topic, as is the case for anything related to climate change, impinges on the perceived vested interests of powerful individuals and corporations. To protect their interests, economically powerful entities such as Exxon-Mobil have joined the communication effort with their own teams of so-called "scientists" who generate press releases and stories all designed to raise doubt about the science.[15] Many journalists, ill-equipped to discern scientifically shaky claims, incorporate into their own stories the material fed to them by these professional deniers.

The average interested *everyperson*, reading the news media, would be hard pressed to figure out what is happening to coral reefs or the real significance of reported items. Paradoxically, the media cover the degradation of coral reefs extensively. They report each new bleaching or disease outbreak, each conference, significant paper, or new projection of likely future trends. But the result is a confusing, contradictory mix of more-or-less garbled messages.

I confirmed this conclusion to my own satisfaction in mid-2018 by doing a quick survey of media reports after the appearance of an important paper published in *Nature*. This was just a quick skim using Google, hunting out interesting articles much as would anyone attempting to keep up with what the media were saying. I gave preference to well-established print media and reputedly authoritative news sources, including some web-only ones. Here's what I found:

On April 18, 2018, *Nature* published the latest in a series of papers on the massive 2016 bleaching of the Great Barrier Reef, by a team led by Terry Hughes of James Cook University.[16] This one focused on the pattern of coral mortality at different levels of heating, differential responses to warming by different coral taxa, and the longer-term consequences of the mass bleaching in terms of ecosystem structure and function. The authors showed that some heat-sensitive taxa died from the direct effects of warming, that others died sometime after loss of their symbionts due to the physiological impairment that resulted, and that still others died still later due to secondary mortality factors such as diseases facilitated by the deteriorated condition of the corals. They painted a bleak future in which the Great Barrier Reef will substantially reorganize itself (in terms of species composition and relative abundance, and of ecological processes) in an altered, warmer world, and concluded by suggesting that what is happening on reefs could be "a harbinger of further radical shifts in the condition and dynamics of all ecosystems, reinforcing the need for risk assessment of ecosystem collapse, especially if global action on climate change fails to limit warming to 1.5–2°C above the pre-industrial base-line."

Lead author Terry Hughes had disseminated a press release embargoed until the day the article appeared on the *Nature* website. The same day, *Science Daily* put the press release up verbatim on its site, and the *Atlantic* printed an article by Robinson Meyer that depended heavily on the press release.[17] Meyer's piece is scientifically accurate and captures the main points of the *Nature* article, but the science is so wrapped in poetic metaphors that I think many less-informed readers would come away confused. It begins, "Once upon a time, there was a city so dazzling and kaleidoscopic, so braided and water-rimmed, that it was often compared to a single living body. It clustered around a glimmering emerald spine, which astronauts could glimpse from orbit. It hid warm nooks and crannies, each a nursery for new life. It opened into radiant, iris-colored avenues, which tourists crossed oceans to see. The city was, the experts declared, the planet's largest living structure." A good thing that paragraph follows the simple title: "Since 2016, Half of All Coral in the Great Barrier Reef Has Died."

The same day, Peter Hannam, science reporter at the *Sydney Morning*

Herald, also made use of the press release in reporting.[18] He picked up on Hughes's casual reference, when interviewed, to the reef being "cooked," and this became the first word of his title: "'Cooked': Study Finds Great Barrier Reef Transformed by Mass Bleaching." That phrase got picked up by other print media and on Twitter.

The next day, things began to go downhill. Graham Lloyd, environment editor at the *Australian*—flagship of Rupert Murdoch's News Corp—drew on the press release but put his story under the title "Not All Scientists Agree on Cause of Great Barrier Reef Damage."[19] He quoted Jochen Kaempf, at Flinders University, in Adelaide, as saying, "The claimed link between the 2016 heatwave and global warming has no scientific basis." This statement was taken out of context and concerned the tiny detail of whether the anomalously warm sea surface temperatures were a direct consequence of climate change. Kaempf protested. The *Cairns Post* picked up on Lloyd's creativity the following day under the heading, "Link between Great Barrier Reef Bleaching and Global Warming 'Has No Scientific Basis': Researcher." Down the rabbit hole we went.

The thread unraveled as journalists' attention was drawn to newer reports dealing with corals. Two articles by Mikhail Matz, University of Texas, and colleagues, appearing in technical journals on April 19 and 24, concerned genetic studies that happened to use a Great Barrier Reef coral species. These resulted in media reports titled, "Corals Can Withstand Another Century of Climate Change" and "First Genetically Engineered Coral Created to Help Save Reefs from Climate Change," distorting the scientific results Matz had reported.[20]

Subsequent headlines I found were "Australia Pledges Millions of Dollars in Bid to Rescue Great Barrier Reef" (*New York Times*, April 29), "Great Barrier Reef's Five Near-Death Experiences Revealed in New Paper" (*Sydney Morning Herald*, May 28), "How Justin Trudeau and Jerry Brown Can Help Save the Great Barrier Reef" (*New Yorker*, May 30), "World's Largest Coral Reef Farm Set for Fujairah" (*Gulf Today*, June 1), and "Coral Decline in Great Barrier Reef 'Unprecedented'" (*Guardian*, June 5). These referred respectively to a preelection funding announcement by the Australian government, a new study of the geological history of the Great Barrier Reef over the past thirty thousand years, the fact that political leaders of Australia, Canada, and California were doing little to reduce production of fossil fuels (the one essential action that could assist coral reefs), a routine announcement of a new business enterprise in the United Arab Emirates to farm corals commercially (the report was not clear on the uses to which the farmed coral would be put), and the release of the annual report from the long-term reef monitoring project run by the Australian Institute for Marine Science, a thirty-year-long record of coral decline on the Great Barrier Reef.[21] None of these directly related to the 2016 bleaching or to the effects of climate change on coral reefs, but if one just scans headlines, they suggest, respectively, that the Australian government has matters in hand, that the GBR has nearly died five times, that the likes of Justin Trudeau and Jerry Brown could save it, that farming of corals is under way in the Middle East, so all is now well, and finally that the deterioration of the GBR is unprecedented.

Given my survey, I'm not surprised that the average *everyperson* might be a bit confused about what is happening to coral reefs. The sequence of headlines bounces us back and forth from despair to optimism; journalists have been seduced by the hype in press releases by the scientists; errors in interpreting the science have been made; and in some cases (*Australian*, *Cairns Post*), there has been a deliberate effort to mislead.

The audience: And then there are the readers, the *everypersons* who, in talking to one another, create public opinion, who vote, who support (or not) government actions on climate change. We, too, must bear part of the blame for the failure of communication, although in our case the blame is tempered. We may be shirking the responsibility to be well informed and contribute effectively as members of our societies, but it is the education available to us and the cultural norms in societies that have left many of us less able than we might be to evaluate the news provided by the media.

Sounding very elitist for a moment, I think most of us in advanced Western societies today share the following limitations: The ability to evaluate news critically is weaker than it should be. The sense that understanding the issues of the day is an important part of citizenship is poorly defined. The capacity to discriminate fact from hypothesis or to spot the logical fallacies in an argument is more limited than it should be. The distinction between belief and fact is poorly recognized. And the idea that there are fundamental truths and absolute impossibilities is increasingly being questioned. Add to these problems the fact that most of us are completing our formal educations with little retained ability to deal with quantitative data or to recognize the difference between linear and exponential patterns of change. It's not surprising that comprehending the scientific complexity inherent in any environmental science story becomes very difficult, even for someone trying hard to do so.

And the coral reef story is complicated because corals and reefs are complicated. Just think of the many special facts underlying any story about degradation of coral reefs: Coral reefs are biogenic rocky masses that are dynamically balanced between rates of calcification by corals and some other reef organisms and rates of reef erosion due to wave action, storms, and action of numerous bioeroding species. Corals and coral reefs are entirely different entities despite the fact that bleaching is a response by corals that has direct consequences for reef degradation. Corals are the major calcifying organisms on coral reefs, but they depend on an intimate symbiosis with minute photosynthesizing dinoflagellates that live within the coral's tissues. Physiological stress, such as that caused by warmer than usual water, breaks down this symbiosis, and without their dinoflagellates the corals are compromised and may die. Coral cover is a standard measurement to quantify the abundance of living coral on a reef, and loss of coral cover is a measure of the extent of coral death caused by a bleaching event or other disturbance. Many factors can degrade coral reefs by reducing their coral cover: excessive warming, severe storms, outbreaks of the crown-of-thorns starfish, numerous coral diseases, siltation, coastal pollution, and sea level change are some of them. These factors can act together or separately and can be differently severe in different locations or at different times. "Death" of a reef is a colloquialism referring to severe reef degradation because a reef is not a single organism capable of dying but a collection of many organisms each of which may die. When many corals on a reef die, it is common to speak of the reef as now "dead"—it has lost substantial coral cover—but it will "recover" if recruitment of new juvenile corals and growth of any corals that survived are able to substantially restore its level of coral cover.

Such facts (this list is incomplete) are part of the unspoken fundamental knowledge possessed by any reef scientist or manager and by many other people, but most of us lack this knowledge and will find media accounts of what is currently happening to coral reefs difficult to interpret. We should not expect the *everyperson* to know the details of coral reef ecology; we all have our own lives to live, responsibilities, and interests beyond becoming informed about reefs. Yet articles in the media are overflowing with facts (the scientist's *and, and, and*) while omitting a lot of fundamental knowledge; this makes it difficult for the reader to get the gist of what is happening.

Poorly equipped to understand the scientific details, buffeted by sensational headlines, whipsawed back and forth between despair and optimism, most readers see the prevailing coral reef message as "reefs are being harmed, scientists are making discoveries, there is concern, but there is also reason for optimism." That is a story that is not particularly interesting, certainly not a story that will keep *everyperson's* attention. If you do not depend directly on a coral reef, it's just another just-so nature story.

My quick Google survey is hardly definitive, but I think it illustrates the problem faced by an interested *everyperson* attempting to understand the coral reef crisis.[22] Communication of the coral reef story can be improved substantially. If anything, too many stories in the media delve deeply into the nitty-gritty of particular scientific studies and too few provide the needed overview and a wider perspective.

♦♦♦

The relative ineffectiveness of media reports and the well-funded denialism campaign designed to cast doubt on the reality of climate change are two reasons why the coral reef story fails to resonate with the public. There are several other reasons. These include our lack of direct, personal familiarity with coral reefs, our need to believe in a predictable universe or at least one in which existential crises are just not possible, our own selfishness and short-termism, and subtle aspects of our cognitive systems that determine how we receive and act on ideas.

As a scientist, I have difficulty recognizing that people do not make most decisions based on a rational analysis of the evidence. Most people do not try and become rational evaluators of the evidence even if a scientist points out to them that they should do so. It is not that we fail to succeed at rational analysis of the evidence; it's that we don't put all that much stock in the value of acting rationally!

As a scientist, I have had to learn that most people do not go through their lives acting as a scientist would. And I have a nagging suspicion that I also do not go through life acting as a scientist would! Apart from the decisions we make in order to satisfy immediate personal needs—everything from scratching an itch to grabbing food when hungry to seeking sexual gratification—our decisions may contain an element of rationality, but they are also driven by a host of other psychological factors that enable our success as social animals. The environmental psychologist Robert Gifford talks about our dragons of inaction.

Based at the University of Victoria, on Vancouver Island, Gifford investigates the psychological aspects contributing to why we think and act the way we do on environmental issues. In a 2011 paper in *American Psychologist*, under the delightful title, "Dragons of Inaction," he discusses seven general types of psychological barrier that together impede our ability to act to remedy climate change.[23] As Gifford puts it, there is a "gap between attitude ('I agree this is the best course of action') and behavior ('but I am not doing it') with regard to environmental problems." This gap is created by these dragons, the psychological barriers. Gifford did not devise these barriers, but he was the first to bring discussion of them together. He identifies twenty-nine barriers and groups them into seven dragons: limited cognition, ideology, comparison with others, sunk costs, discredence, perceived risk, and limited behavior.

By limited cognition, Gifford means that we are famously less rational than we tend to assume. Within this category, he includes ignorance, and the fact that our brains evolved to respond to immediate, personal threats, not to slowly approaching dangers. Among the *everypersons* there is ignorance surrounding the killing of coral reefs by climate change and about what can be done about this. Reducing this ignorance is made difficult by the technical complexity of the subject and by the mixed messages in the media, whether caused by ineffective simplification of complex issues or by deliberate denialist campaigns.

In addition to this ignorance, Gifford's limited cognition includes environmental numbness, uncertainty, judgmental discounting, optimism bias, and perceived behavioral control/self-efficacy. (I cannot resist noting that the ability to complicate, while trying to explain, is not limited to ecologists.) Environmental numbness arises because we are organisms that always tune out perceptions that are less immediately important to us and because we also tune out repeated messages, the no longer new news. The inherent uncertainty in any scientifically grounded message causes us to undervalue it relative to competing, apparently more certain messages because we avoid the risk of changing concepts or actions unnecessarily. As well as failing to respond to uncertain messages, we are poor at estimating risk, tending to discount more than we should those risks that are far away or in the future. In each of these aspects, our evolutionary selection to avoid nearby saber-toothed cats and ignore less pressing problems is obvious, and any message about the bleaching of a coral reef far away is definitely less pressing. Add in optimism bias—the tendency to see the cup as always at least half full—and messages about reef degradation become very weak motivators indeed for the great majority of the audience. Gifford's final component of limited cognition is the sense that my individual actions can have no real effect on such a large problem. *Everypersons* do not understand the links between warming, bleaching, and reef degradation, have difficulty becoming better informed, see the problem as not important enough to act on, and see any actions they might individually undertake as inadequate to the task. Those who craft the coral reef story can present the story more clearly. But without exaggerating and distorting what is known, they cannot make the story appear more important, or a response to the story more urgently needed. Making scientifically unjustified claims of certainty about future events might galvanize action by the public, but at the real risk of blowing up any credibility environmental scientists still possess. Presenting the information accurately and dispassionately leads to an ineffective uptake by the lay public.

The second of Gifford's seven dragons is ideology. He includes here worldviews, suprahuman powers, technosalvation, and system justification. A commitment to free-market capitalism is the worldview least likely to favor action on climate. Formal religious beliefs in an all-powerful deity that cares about individual humans, more secular belief in an all-powerful Nature that will always repair itself, faith in humanity's capacity for technological innovation to solve any problem, and strong desire to not disturb the status quo sociopolitical system are all aspects of ideology. In all these cases, belief systems can provide strong incentives to not act on messages about climate change and reef decline. Putting it more simply, beliefs trump facts.

Comparisons to others is the third dragon. Included here are social comparison, social norms and networks, and perceived inequity. People tend to act in ways that mirror actions of others in their social group. Members of effective social networks perceive and emulate normative actions including altruistic acts; however, when societies are perceived as inequitable, the tendency to behave unselfishly is reduced. Messages about reef degradation being delivered in the hope that they will galvanize concern and action on climate change must take account of the social milieu in which the *everypersons* spend their lives.

Sunk costs, the fourth dragon, can impede action, even if the messages have registered as important. These costs include behavioral as well as financial costs, conflicting values, goals, or aspirations, and attachment to place. Habitual patterns of behavior impede behavioral change simply because they are habits, in the same way that prior investment in particular ways of doing things can impede changes in behavior because change will cost money. Humans rarely find their various individual goals all aligned, and conflicting values, goals, or aspirations can all impede the behavioral change needed to solve environmental problems.

Discredence, the fifth dragon, refers to the tendency to disbelieve or even to be defiantly opposed to those telling the coral reef story if they are not members of the right tribe. Gifford sees trust as essential for listeners to receive messages being delivered, and trust of citizens toward scientists or government officials is easily lost. A more aggressive level of opposition comes in the form of denial. Sometimes denial is simply a way of rationalizing a desire not to comply with or support environmental action, but sometimes denial is motivated by fear and becomes a way of blocking out bad news. (This denial by recipients of messages is entirely distinct from denialism practiced by individuals or corporations deliberately seeking to sow confusion about the messages being sent.)

In July 2018, a U.S. study confirmed that the majority of Republicans, as well as the majority of Independents and Democrats, accept the reality of human-caused climate change. That good news came with a caveat. When people were asked to evaluate a climate policy (either a cap-and-trade or a tax proposal to reduce carbon emissions) ostensibly being considered by the legislature, tribe or beliefs trumped logic. Democrats disliked Republican policy and Republicans disliked Democratic policy even when the policy description was identical.[24] Discredence at work, clear and simple.

Conservation science has long recognized the importance of respected local leaders in encouraging the adoption by a community of novel conservation actions. A respected older fisherman is far more persuasive when articulating the benefits of establishing a no-fishing reserve on a reef as a way to sustain the local fishery. Far better the community hears it from the respected elder than from a youngish, foreign-looking stranger who flew in to talk to them one Friday. The same clearly applies to American members of political parties and, I suspect, to all advanced societies. Better we hear the novel proposal from someone just like us. Otherwise, on what basis do we believe the message?

Gifford's sixth dragon is perceived risk. This is the risk one takes on by changing behavior in order to respond to a message about climate change or reef degradation—the risk of acting, rather than the risk incurred by ignoring the problem. Again, Gifford sees several different types of risk in acting: functional, physical, financial, social, psychological, and temporal.[25]

Gifford's final dragon is limited behavior. Today, whether responding to the coral reef story or to other climate change messages, many people are engaged to some degree in actions to reduce climate change. Some people are more engaged than others, yet most of us could do more than we do. As well as buying a plug-in hybrid car, I could become a vegetarian and put solar panels on my roof (among many other choices), but I have not done so. In this way, I have limited my behavioral change. Gifford divides limited behavior into tokenism and rebound. Tokenism involves the adoption of easier, less costly, or less disruptive changes to behavior while ignoring other possible changes. Most of us are guilty of this to some degree. Rebound occurs when, having made a behavioral choice in favor of reducing emissions, we slack off, perhaps bringing our carbon footprints back to where they were before we acted. Taken together, tokenism and rebound act to limit what each of us might do in response to the climate emergency. In essence, we are accepting that there needs to be a reduction in emissions of carbon dioxide, and we make modest attempts in that direction; then we move on to other things (because it's not healthy to obsess constantly about climate change), perhaps increasing our emissions in the process.

Gifford argues that, if the goal in delivering environmental messages is to effect change in behavior and to build support for action to curb climate change, these seven dragons of inaction must be dealt with. Otherwise they gang up and their fiery breath impedes both message receipt and any hoped-for action in response. He notes that some of these dragons may be far more important than others, and like all good academics, he calls for more research. But he also makes clear that how individuals respond to messages about issues like climate change depends as much on social milieu, emotion, and motivation as it does on the nature of the message sent. If by telling our stories, scientists seek to inspire people to act on climate change, we need to find ways of slaying these seven dragons.

♦♦♦

Climate change, coral reef degradation, and the wider environmental crisis inevitably become a political problem the moment we move from describing what is happening to talking about changes to our behavior to remedy, repair, or avoid such problems in the future. Even if scientists and journalists present the facts clearly—which we usually don't—the discussion will not be entirely rational, a dispassionate examination of the facts. Neither political systems nor people work like that, and coral reef scientists have been slow to learn this.

For a reef scientist like me, it is now obvious that we are not going to be able to retain functional coral reefs on this planet unless we limit climate change substantially. But it's also obvious that solving climate change is going to require a monumental global political effort—the largest such effort ever attempted, at a time when we are still infants in learning how to behave as an effective global community. We have made halting progress using the politically weak structures we have in place, and the pace has been distressing to those like me who recognize the seriousness of the problem and want to see solutions. Political action requires consensus (except in systems run by autocrats), and global political action is challenging—especially in a world of nations with different worldviews and modes of governance, including some autocrats. It is not the job of the scientist, or other technical expert, to solve the problems of how to effect global change, but that expert does have an obligation to attempt to provide sound information to people who have the skills to navigate the passageways, tunnels, and private rooms that constitute the path toward effective political action. Finding the most effective ways to influence political leaders, *including finding ways to encourage strong, coordinated pressure from the everyperson constituents of those leaders*, and finding effective ways to reward positive achievements while shaming inaction or action in inappropriate directions, must be a high priority for those who want climate change brought under control. Gifford's seven dragons stand before each of us.

Two parts of the coral reef story deserve wide promulgation if we are to mobilize effort on our environmental crisis. The first concerns our current understanding of the immense value of coral reefs biologically, economically, and aesthetically. It deserves more than the reporting of some facts and figures about numbers of dependent people, contribution to gross domestic product, and vague waffle about solace for the soul. I think the case can be made that we have an obligation to humanity, and a moral obligation to the planet, to act to minimize our unintended negative impacts on coral reef systems, but making this case in ways that will cause it to be taken up enthusiastically will not be easy. The second part of the story concerns the canary connection between climate effects on coral reefs and the concern of many environmental scientists that human activities have begun to shift Earth beyond the planetary boundaries that define a Holocene-like environment. For me, this connection is ultimately what makes the coral reef story deeply troubling, because a non-Holocene world will deal harshly with the continuance of our civilization. We have it within our power to address the size of our footprint on this planet, but we are acting far too slowly.

A concerted effort to convey both these parts of the coral reef story to the *everypersons*, using effective storytelling techniques, could be far more effective in raising awareness and concern about the decline of coral reefs and in building understanding of the perils we are creating for ourselves around the world. Along the way, we must also name and confront the dragons. The societal changes needed are unlikely to occur without this.

ELEVEN

We've Left the Holocene

It was early afternoon of Monday, August 29, 2016, in Cape Town, South Africa, when Colin Waters gave his report to the Thirty-Fifth Annual International Geological Congress. As secretary of the Anthropocene Working Group, he presented the findings of many years of effort and the nearly unanimous recommendation that the Anthropocene be formally recognized as the most recent epoch in the geological record. It would follow the Holocene, which would now be ended. Thirty-four of thirty-five working group members (one abstained) agreed that the Anthropocene was real, and thirty (three against, two abstaining) agreed that it should be formally recognized.[1]

Waters was talking about adding a new epoch to the history of Earth, a new chapter in the history of this corner of the universe. As is appropriate under such circumstances, the global geological community proceeds cautiously, and there will be a couple more years before things become official, but in 2020 it looks as if the scientists are zeroing in on formal recognition, with 1950 or thereabouts as the start date and a global plutonium spike as the primary geological marker of its start. Declaring that time as the start of a new chapter in the planet's history means formal recognition by the geoscience community that significant changes have occurred. Those changes were not just the H-bomb testing that led to the sudden spike in abundance of plutonium but the rapid acceleration in the rate at which a range of human activities has been disturbing the planet. That acceleration began in the early postwar period as the global economy exploded; it has been called the Great Acceleration by environmental scientists interested in human impacts.

The Anthropocene is here, and human influences are radically altering the planetary system in ways that did not occur in the past. This disturbance or change in the status of the planetary system, caused by the activities of a single species of life (*Homo sapiens*), is the Anthropocene's defining feature. Nothing remotely like this has ever occurred in the 4.5-billion-year history of Earth until now.[2] The Anthropocene is real whether or not the geological community eventually decides to recognize it formally. Formal recognition by the International Commission on Stratigraphy will bring precision to the term *Anthropocene* and will confirm that it is not just a few leftist environmentalists who use this word.

The Holocene, in which we lived until now, was also marked by the growing might of the human population, notably by the development of agriculture and the substantial changes to the terrestrial landscape that we wrought as our agricultural activities expanded. It has been a brief epoch, commencing just 11,500 years ago as the most recent Pleistocene glaciers retreated; it has also been a surprisingly benign one apart from our agriculture-driven rearrangement of terrestrial landscapes. Who knows if the Anthropocene will be similarly brief or to what it will lead. It's not official yet, but welcome to the Anthropocene: we may be in for a wild ride.

♦♦♦

Our impacts on coral reefs have been profound, and reefs will not emerge from this century resembling what they were like when I began to study them. In fact, my best bet is that we will probably have lost them well before 2100, if one defines a coral reef as a reef with substantial amounts of living coral present. Although that's my best bet, I am optimistic that reefs can remain a part of our world if we grapple successfully with the damage we are doing to the climate and if we also address our more local assaults on reefs themselves. They won't resemble coral reefs of the 1950s or 1960s, but they could still be functional, relatively high diversity ecosystems, providing a range of services to coastal communities. To get there, we need to look more critically at how we manage coral reefs themselves, but we also have to find a way to transition our global economy to one that is far more sustainable than our fossil-fuel-intensive economy of today. Let's consider this transition challenge.

In chapter 10, I considered reasons for our failure to grasp the message that coral reefs were being severely degraded by our actions, and especially for our failure to see the wider importance of this message and the need to act. Those reasons had as much or more to do with how humans and human societies think, decide, and behave as they did with how the coral reef story was being told and how active denialism was muddying that message. When it comes to how we might coordinate the massive, sustained actions required to correct past behavior, I think we face even greater impediments.

Many people see the coral reef story as one that calls on us to act promptly to sustain and conserve coral reefs. I prefer to view it as a small handful of threads in a complex weave of seemingly separate stories that all point to the need to act in ways that will collectively repair our planet. In my view, we will not sustain or conserve coral reefs without substantially halting the emissions of greenhouse gases that have produced the warmer ocean. Looking only at coral reefs, thinking narrowly about reef repair, is far too narrow a perspective to be successful overall.[3] I also think that the task in front of us is so immense that it will take the combined impact of all the stories to get us moving and keep us focused—stories of melting permafrost, disappearing rain forests, the growing water crisis, and the oversized human footprint. Most of all, we need a narrative that encapsulates these disparate stories, because too few of us understand, at present, that they are interlinked and that our future quality of life is intrinsically intertwined with maintaining a world of functional ecosystems that sustain other species as well as our own.

When we bring all the stories together, the true immensity of our problem becomes apparent, and this creates its own complication. Confronted with such an immense problem, many people will simply shut down. Several of Gifford's dragons will belch out fire and brimstone—the task is too big, I cannot believe it's this bad, the world is too dependable for this to be happening. Meanwhile, denialist campaigns will nibble away at details, impeding particular responses that might be detrimental to certain interests, and even those striving to change our behavior will persist in assuming that conventional political compromise—the way we usually agree on collaborative tasks—is an appropriate way forward.[4] Such attitudes will slow us down and weaken our eventual effort.

Thus, we have a nasty conundrum: to truly understand the task confronting us, we need to grasp the full consequences of present human actions—we need to hear and comprehend all the stories—but we also need to see the path forward as achievable. That combined story directs us to undertake a monumental, globally collaborative, if not globally integrated, suite of actions sustained over many decades if we are to emerge in a quasi-Holocene future. This monumental task is way beyond anything humanity has attempted in the past, and our global political structures seem insufficient to achieve this task. What can we do other than huddle down in a fetal position and wait for disaster?

The narrative that encapsulates all the smaller environmental stories needs to be recast in a form that will allow it to be taken to heart. Until now, discussions of solutions to climate change have focused on technical details. We must reduce emissions of greenhouse gases by the equivalent of x metric tons of carbon dioxide. We must establish new carbon sinks capable of sequestering y metric tons of carbon. We must decarbonize our economy totally. We must do all these things on a specific schedule in order to achieve an average global temperature increase of less than 2°C (or 1.5°C if possible). Presented that way, the problem can be understood by technical experts, but not by the vast majority of humans, including the politicians and policy experts who will have to lead the way in securing collaboration across nations.

I'm prompted to reflect on yet another narrative, one with some relation to coral reefs because of where it takes place. This is the story of Hōkūle'a (fig. 24).[5] Polynesia is a vast, endless, beautiful, 25-million-square-kilometer ocean with a scattering of tiny islands and reefs. Over two thousand years, the Polynesians explored this ocean, successfully colonized every habitable speck of land except for Pitcairn Island, traveled, and traded. They were sailing from one tiny island to another across the vastness at a time when European and Chinese explorers tended to stay close to continental shores. The Polynesians' double-hulled ocean-sailing canoes had only the winds to power them, and navigation was done without compass, sextant, or any com-

Figure 24. Hōkūle'a, the first modern ocean-sailing Polynesian canoe. She has voyaged extensively since her launching in 1975, always using traditional Polynesian wayfinding. Photo by Kaipo Ki'aha, © Polynesian Voyaging Society and 'Ōiwi Television Network.

parable instruments. All navigation took place within the mind of the navigator, or wayfinder, the most important person on board.

Over time, the impetus to trade diminished, and by April 13, 1769, when James Cook arrived in Tahiti, there had been no long-distance oceanic voyages for several generations and skills were being lost. Even so, the Tahitians gave Cook valuable information on the directions and distances of various other island groups, and the priest Tupaia guided Cook to New Zealand.

In the 1970s, amid a general reawakening of interest in their culture, Hawaiians decided to try to rediscover ancient Polynesian wayfinding. Perhaps this interest was heightened by the clash between the old stories, handed down verbally, and the generally accepted (perhaps racist) view of anthropologists that the Pacific had been colonized by accident, by fishermen blown off course in storms, or even by South Americans drifting westward (again blown off course) on balsa rafts.

Hōkūle'a was launched in 1975, as faithful as possible a replica of an ancient, ocean-sailing canoe. It was double-hulled, 18.6 meters in length and 6 meters wide, built of traditional materials using traditional methods. It contained no metal of any kind and carried no instruments. While the ship was under construction, a small group under the guidance of Mau Piailug, a traditional navigator recruited from tiny Satawal, far to the west in Micronesia, devoted themselves to exploring the ancient stories, talking to elders, and trial and error working out how Polynesians might have navigated.[6] In 1976, Mau Piailug guided Hōkūle'a to Tahiti—4,400 kilometers across an empty ocean and into a hemisphere he had never sailed before. He continued working with the Hawaiians, particularly Nainoa Thompson, who became the first modern Hawaiian navigator. There have been several subsequent canoes and voyages, and four other Hawaiians have joined Thompson as Master Navigators. Each navigator uses a detailed knowledge of the movements of the sun, moon, stars, and planets, the patterns in the ocean swell, details of the creatures passing by, cloud formations, and the smell of the air to continually adjust a mind map that tells him (or her) where he is and the direction to sail to reach his goal. In 2017, Hōkūle'a completed a three-year circumnavigation of the planet navigating throughout using Polynesian wayfinding. Validating their cultural beliefs, those Hawaiians have succeeded in reviving a magnificent body of knowledge and a way of life. They also have inspired similar efforts throughout Polynesia.

I think we now find ourselves in very much the same situation in which those Hawaiians found themselves forty years ago. They believed the traditional stories of long-distance voyaging, knew it must be possible, but they had to figure out how Polynesian wayfinders did it and then apply those methods to once again sail successfully across vast stretches of open ocean. We know we have an immensely complex problem, and we understand what has caused it. We can visualize a number of possible procedures that should help us solve that problem, but it is a much bigger, more complicated problem than any we have had to solve before. We have got to learn how to navigate our planet into those safe waters where a quasi-Holocene biosphere will flourish, sustaining our civilization while also nurturing the natural ecosystems that sustain our lives. It's not a set of separate environmental problems, each to be fixed one by one, but a monumental voyage that has to be taken for our collective well-being. It's a voyage in which we will be guided, not by stars, but by changing concentrations of greenhouse gases, extent of Arctic ice, abundances of living corals on reefs, and many other environmental indicators. These will tell us if we are moving in the right direction. Portraying our problem as a challenge, and that challenge as a project to use our very considerable power to steer our planet toward safe waters, creates an image that is positive, somewhat heroic, and one that people will be able to wrap their hearts around as well as their heads. That will help us make the difficult decisions and take the actions we need to take to build a quasi-Holocene future.

♦♦♦

Rather than a technically complicated problem, we now have a bold quest before us. Are we ready finally to embark? Not yet, I fear. Four other essentials remain before we cast off. We need broad acceptance of the need to embark upon this journey and the essential correctness of so doing. We need confidence in our collective ability to act together to achieve difficult tasks. We need faith that our scientific understanding and technological prowess are sufficient to steer the planet to a quasi-Holocene future. Along the way, we will need the kind of leadership that is not often seen, because we will need a sustained, long-term focus on distant goals. Such sustained effort demands inspirational leadership as well as a noble venture worth the commitment.

That's all! As well as stories from reefs and elsewhere showing the damage we are doing and an inspiring goal before us, we need *acceptance* of the urgent need to act, *confidence* in our collective abilities, *faith* in the power of our science and technology to serve our needs, and sustained, inspired *leadership*. Just possibly, if we nurture a desire to keep wondrous ecosystems like coral reefs with us on our journey, that will provide the extra kick to get us on our way. I think all these things are at hand or nearly within our grasp. (On my good days, I really am an optimist.)

As I write, in mid-2020, the need to act on climate change seems at last to be gaining significant traction around the world. COVID-19, now raging, has slowed action on climate, but there are many voices arguing that the opportunity to recast economies in less carbon-intensive ways as economies recover is one tiny bright spot in the dark cloud of this pandemic. Even in the United States a clear and growing majority of people recognizes that climate change is real and due to human activities, and we are past the time when politicians can routinely claim that they do not accept this reality.[7]

This shift in attitudes has occurred as evidence mounts of changes in climate that have already occurred and as our experience of severe weather grows. Severe wildfires, intense storms, and major floods, all of which seem to be worse than anyone remembers, are increasingly in the news. Ironically, it's the severe weather (only sometimes due to climate change), rather than demonstrable evidence of climate change, that is more compelling.[8]

Projections of the dire consequences of continuing as we have been, and warming the planet further, become much more real to people as they confront increasingly damaging weather events that are happening right now. Canadians are not yet willing to pay very much for climate action, but they want their governments to do something. Australians, who in 2019–2020 suffered through the most severe fire season in history and witnessed another severe bleaching of the Great Barrier Reef, may already be at a tipping point, demanding that their government finally do something about their carbon dioxide emissions. That's a start that we can work with.

We can build *confidence* in our collective ability to tackle this task, and *faith* in the capacity of our science to help us, by reminding each other of three notable successes in tackling complex environmental problems. These three battles repaired the environmental damage being done by DDT, acid rain, and stratospheric ozone depletion. None were quick, decisive battles, and they provide important lessons for the struggles we now face, but all three are global battles we won using existing science and established mechanisms for international diplomacy.

The DDT story as usually told notes that this synthetic insecticide was widely used by the U.S. military during World War II to control diseases such as typhus and malaria and became available after that war for agricultural and domestic use. It was the new miracle cure for a host of conditions; as a child I remember the Flit gun kept ready to ward off cockroaches or other critters that invaded our home. Use of DDT exploded in agriculture and domestically, but in 1962 Rachel Carson wrote *Silent Spring*, documenting our chemical pollution of our environment and singling out DDT for its health and environmental risks. The United States banned DDT for all except carefully controlled antimalarial use in 1972, and it subsequently has been largely eliminated from use around the world.[9]

The real story is a bit more complex. It's a story of large chemical corporations, striving to ensure continued sales of their product following the war, aggressively promoting this new wonder product, having ramped up production during the war years. Predictably, these companies fought strenuously to continue to promote DDT even as evidence of unintended environmental and medical impacts grew in the 1950s and 1960s. The eventual banning in the United States in 1972, by the then very young Environmental Protection Agency, was forced by court actions initiated by the newly formed, nonprofit Environmental Defense Fund—a novel approach for environmental NGOs at that time. Internationally, the effort to eliminate use of DDT was complicated by its evident efficacy in managing diseases such as malaria, and it was not until 2004, when the Stockholm Convention came into effect, that DDT use was well regulated. Concentrations of DDT residues in wildlife and humans are slowly falling.[10]

The story of acid rain might have begun when I was working on my M.Sc. project out of the University of Toronto, except that I failed to see a problem in front of my nose. It was 1962, and I was tasked with finding out why a rare, beautiful fish, found in just two small lakes in the Temagami region of northern Ontario, was apparently going extinct.[11] This was a land of pristine headwater lakes, isolated, undeveloped, and several days by canoe from any possible sources of damage; why should the fish be disappearing? Being a dutiful aquatic biologist, I collected routine water quality information and noted that the waters of these lakes were rather acid (pH of 5.4 to 5.5 for White Pine Lake). End of story.

Although low pH can disrupt reproduction in trout, in those pre–*Silent Spring* days, neither I nor the talented professionals mentoring my work ever suspected that pH in these lakes was being made low, and becoming lower year by year, because of human-caused air pollution. It was nearly a decade before Ontario scientists made the connection.

In 1967, Richard Beamish had commenced a Ph.D. program at University of Toronto on the biology of a common freshwater fish, the white sucker, *Catostomus commersonii*, and particularly on why they varied greatly in their growth rates and longevity from lake to lake. In the La Cloche region of small lakes north of Lake Huron, where fish populations of many species were dwindling, lakes were becoming more acid. During five years from 1967 to 1971, pH fell an average of 0.21 units per year across twenty-two lakes sampled by Beamish, and in 1971, pH in fifteen of these lakes was between 4.0 and 4.9. Like my lakes, these were also far from human settlement and industry. Scandinavian researchers had observed similar acidification of lakes and had attributed this to airborne sulfur pollution from the industrial regions of central and western Europe. Power plants and smelters across Illinois, Michigan, Indiana, Ohio, and Ontario were dumping sulfur dioxide and other pollutants into the atmosphere. The sulfur dioxide would interact with water vapor to form sulfuric acid with measurably acid rain the result. Rain in the vicinity of Beamish's lakes had pH of 3.6 to 5.5 in 1969–1971, and snow samples taken in winter of 1970 were even more acidic (pH of 2.9 to 3.5).[12]

The U.S.–Canada Air Quality Agreement, signed in 1991 and expanded over the years to address other types of transboundary air pollution, has been responsible for reducing pollution by sulfur dioxide by more than two-thirds from 1990 levels, bringing acid rain under control in both countries.[13] It did so by requiring each government to set mandatory targets for pollution reduction, grant industries permits for a certain level of pollutants emitted, and then leave it up to industry to find ways to cut emissions and sell off their unused permits to others. This approach is now called cap-and-trade. This treaty brought acid precipitation under control in North America.

In October 1984, the amount of ozone measured above Antarctica was reported to have fallen below 200 DU (Dobson units); it had been at or above 280 DU before 1970 and had been falling since the mid-1970s.[14] Stratospheric ozone, although incredibly rare—typically ten or fewer molecules per million molecules of all atmospheric gases combined—plays a vital role in protecting living tissues from genetic damage caused by ultraviolet solar radiation. I was living in Australia at the time, and the thinning ozone layer was big news Down Under. Over a few short years, a nation of bronzed, blond-haired, outdoor-loving people began wearing sun hats, and a streak of white zinc sunblock on the nose became a new fashion statement for the surfing crowd.

Atmospheric concentrations of certain chlorofluorocarbons (CFCs) had been increasing at the same time that ozone concentrations were falling. CFCs are chemicals invented and manufactured for use as refrigerants, aerosols, foaming agents, and solvents; one widespread brand is Freon. First manufactured in the 1930s, they were in widespread use by the 1970s. However, by 1974, the possibility was recognized that they would become aggressive destroyers of ozone if released into the atmosphere.[15] Clearly, that had happened.

So much for the science. We had all conveniently forgotten that refrigerators leak, and aerosols are sprayed, and relatively lightweight CFCs float upward. We had a culprit (CFC) well before the damage to the ozone layer was detected. We also could see clearly what was needed. We had to stop using these chemicals because we could not afford to lose the ozone layer. We began the effort to phase out CFCs globally even before scientists reported thinning of the ozone layer!

In 1981, UNEP (U.N. Environment Programme) began drafting an international convention on stratospheric ozone protection. That agreement, the Vienna Convention, was opened for signing in March 1985; ratifying countries agreed to cooperate in research and development and to commit to not-yet-specified actions to slow or halt ozone thinning. In just two further years, the science had progressed sufficiently that UNEP was able to launch the Montreal Protocol, providing the initial targets for action to phase out manufacture and use of CFCs and some related chemicals. Drafting the Montreal Protocol had required just nine months. It was opened for signing in September 1987 and went into effect on January 1, 1989. Over the past thirty years, all 197 U.N. countries have become parties to it. Its requirements have been strengthened several times, and the HFCs (hydrofluorocarbons) that were originally used to replace CFCs are now in turn being phased out because of their effects on climate. By late 2018, media reported that the ozone layer was rebuilding. Ozone concentration over the Arctic is likely to be largely restored by 2030, and at current pace the depletion over the Antarctic will likely be fully repaired by the 2060s.[16]

DDT, acid rain, ozone depletion—we really do have the capacity to come together in multinational efforts to correct our environmental missteps. But we do not always move quickly, and success is not always 100 percent. It took ten years after the publication of *Silent Spring* to shut down use of DDT in the United States. Opposition by manufacturers who were exporting DDT throughout the world delayed negotiation of the Stockholm Convention a further twenty-five years until 1997. The wider story of persistent organic pollutants (POPs, of which DDT is one) is an ongoing race between manufacturers who create a continuous stream of new (toxic) wonder products and regulators who seek to control or ban use of each new chemical in turn because of the unintended environmental damage they do.

After the irrefutable evidence of lake acidification provided by Beamish, it took Canada and the United States seventeen years to reach agreement on acid rain, again because of resistance from industry facing significant costs.[17] In this instance, Canada led by admitting that it was responsible for half its own, and some U.S. acid pollution, and by building an internal agreement among industry, provinces, and the federal government to cut emissions of sulfur dioxide by 50 percent. With that internal agreement in place, Canada then "encouraged" the United States to also commit to 50 percent.[18] Forging the U.S.–Canada Air Quality Agreement became a textbook example of the cap-and-trade principle and of how to build effective multinational coalitions to achieve shared environmental goals.

Deliberations within UNEP on CFCs took twelve years before the Montreal Protocol came into effect in 1989, still by far the most rapid agreement of the three. Everyone agreed on what needed to be done, and CFCs were already out of patent protection; their manufacturers had newer, more ozone-friendly chemicals that could substitute for them and did not fight the regulations.[19] Lessons learned in the Vienna Convention–Montreal Protocol process are now being used in efforts to tackle climate change. These include the value of *having scientists and politicians work together* as science is extended and policy developed and the idea of *common but differentiated responsibility* (which made it possible to engage developing countries successfully by offering financial assistance and extended time for them to complete cessation of use of CFCs). The idea of *common targets that might be altered as the science developed* was also central in all three of these success stories.

♦♦♦

If I knew the secret of how to provide *inspired leadership* to achieve environmental goals, I'd be writing a very different book. The world of today seems chaotic and alarming; effective leadership to deal with major global problems is not apparent, but I suspect that at least some of the leadership we badly need will come from quite unexpected places. I believe, and fervently hope, that it will come.

To that end, let's say no more about leadership. Let's instead focus on confidence, faith, and acceptance, optimistic that needed leadership will appear as we embark on our grand quest to steer our planet wisely. To solve the environmental crisis we have created, we must use our considerable power to right past errors and move our planet toward a quasi-Holocene state. For this immense task, we must have confidence in our ability as humans, to work together, over sustained periods, to achieve common goals; and faith that our science and technology is up to the task confronting it. We've more work to do on acceptance, but there is already ample evidence to justify that confidence and sustain that faith. We need to talk up that evidence at a time when fake news trending on social media risks turning far too many of us into intellectual zombies. As we start this great journey, we should strive to keep coral reefs in view; they will tell us if we are sailing in the right direction.

TWELVE

Coral Reefs in the Anthropocene

There is something magical about diving on a coral reef. Perhaps it's the warm water, which allows diving with minimal clothing—the thinnest dive skin or perhaps just a Speedo. I vividly remember my first dive in New Hampshire, encased head to toe in a full quarter-inch neoprene suit, hood, and gloves. I found it claustrophobic, and the relatively low light levels made it seem even colder than it was. Then there was that other New Hampshire dive when I was helping one of my students by holding down one end of his transect tape. We were relatively shallow, and I found I could hold the tape satisfactorily against the substratum with the tip of my fin while keeping my body vertical in the water with my head and all vital organs just above the thermocline. I think that was my last cold-water dive. But on a reef, only lightly clothed, one can begin to feel like an aquatic creature. Indeed, during my earliest days in Australia, when dive gear was technically simpler—a tank on a simple backpack, separate weight belt, single hose and reg, and an optional buoyancy compensator that was only ever inflated in an emergency—the reef diver could be delightfully unencumbered. In any event, the less equipment, the better you can imagine that you've become one with the fishes.

The usually clear water and the vivid colors under sunlit skies also help build the magic. Those dives in New Hampshire were experienced in tones of gray (and I'll admit to having made some reef dives under poor weather conditions that were grayish in tone). The magical topography, at all spatial scales, adds to the otherworldliness of the reef dive experience. I have always lived in awe of those divers who began their underwater lives in a Canadian lake or on the New Hampshire or Newfoundland shore—never mind those abandoned quarries that seem to litter the Midwest. Why would anyone bother? Wrapping yourself up in rubber and submerging yourself in dark, cold water with scarcely anything to see: to me that is a special form of masochism. Yet, while I do not usually admire masochism, I do admire the fortitude of those who learned to dive under such conditions. Not for me!

I still recall the first time I put a tank on my back and descended beneath the surface. It was during my first September in Hawai'i. Bill Gosline, curmudgeon with a nurturing, sensitive side he tried to keep well hidden, believed that the obvious thing to do with students who did not already dive was to send them underwater for a five-minute taster. So, on an early field trip to Hanauma Bay as part of our fish biology course, he took half a dozen of us new graduate students aside, with his teaching assistant to help, and one at a time we were blessed with five minutes on scuba. "Remember to keep breathing, especially when ascending, for god's sake." I say "blessed" because the tank and harness were placed over my head and onto my back in much the same way a doctoral hood would be placed in a convocation a few years later. But I also say "blessed" because those five minutes swimming freely yet able to breathe left me feeling that I had gone to some sort of heaven. I was over a reef, surrounded by fish, and I did not have to pause and rise to the surface for my next breath. After I was forced back to the surface so the next novice could have her inspirational moment, I remember Gosline's words: "Now go get a scuba course."

I also remember the first time I snorkeled over a reef after getting contact lenses. My eyes were not terrible, but I used glasses to drive and to see the blackboard. Still, I did not spring for the extra money for a prescription mask—available only as an after-market modification, they were unreliable and expensive. I was a graduate student living on a teaching assistant salary in Honolulu. So, I happily observed an uncorrected underwater world for my first eight months in Hawai'i, until I finally saved up the funds for contacts, and became sufficiently confident about wearing them that I ventured into the water, again at Hanauma Bay, while wearing my lenses. Suddenly the fish had fins with sharp edges. And the corals they passed had sharp edges, too. I was flabbergasted, because I had simply assumed that water, being liquid, would naturally soften edges of everything I looked at. But that was not the case, and with my lenses in place, I found that underwater world more marvelous than I had dared to imagine.[1]

Beyond the immensely pleasurable experience of weightlessly drifting over and around an awesomely majestic seascape, watching myriad creatures of differing shape, size, texture, and hue going about their day, working, and playing, a dive to a reef brings the person who knows something about coral reefs to a truly magical place, a place of fantastic otherworldliness. It grieves me that reefs are becoming less magical as we degrade and simplify them, and deep down I fear that we are almost certainly too late to be able to retain much of their glory into the future.

And yet, if we can build our will to act sufficiently quickly, we should have a fighting chance of keeping some of them with us into the second half of the century and perhaps beyond. As I complete this manuscript, I am encouraged to see coral reef ecologists beginning to discuss the need to assist coral reefs to maintain their ecological functionality even as they are changed by the pressures of the Anthropocene.[2]

In these discussions there is evidence that reef ecologists recognize we have a lot to learn concerning the links between ecosystem structure or organization on one hand, and ecosystem function on the other.[3] We need to begin by acknowledging that we cannot bring coral reefs back to the way they were when I commenced my career. Nor to the way they were even earlier, when pioneers like Hans Hass or Conrad Limbaugh were using scuba and writing about them in the 1950s, or earlier still, the 1920s, when members of the Royal Society Great Barrier Reef Expedition were taking some preliminary looks at subtidal coral reefs using a primitive hardhat diving suit powered by an unreliable compressor on the surface. The environmental conditions that nurtured coral reefs in the mid-1960s, in the 1950s, or in the 1920s, are gone, and we have no way of recapturing them. However, if we steer our planet successfully, we will move in directions favoring retention of those coral reef functions that can be retained, and development of new or replacement functionality that becomes possible as the Anthropocene progresses.

The discussions now commencing among reef scientists and managers reveal an encouraging and substantial change in the perspective that has traditionally guided reef conservation. That do-no-evil perspective saw conservation as a process of reducing identified local impacts on reefs and allowing them to repair themselves. That approach is no longer sufficient, if it ever was, because the reefs that used to be cannot survive, let alone rebuild themselves, in this warmer, more acidic world. Reef ecologists are beginning to recognize that coral reefs are essentially socioecological systems with structure and function driven by societal pressures as much as by physical and biological processes.[4] Reef scientists and managers are now developing a willingness to entertain new possibilities for how complex, tropical marine ecosystems can be assembled around those calcifiers that are able to persist. With a little luck, and some concerted action on climate change—some serious effort to steer our planet—these new possibilities may still deserve the name *coral reefs*.

Such rethinking is not only needed in coral reef management, of course. The Anthropocene is forcing change on most ecosystems, and a backward-looking effort to conserve, by preserving the past, is certain to fail in these other ecosystems, just as it will fail on coral reefs. We need to learn to conserve by looking at ways to shift ecosystems in favorable new directions while helping them avoid unfavorable paths.[5]

Rewilding, in which we give up our control over significant portions of the land and sea, will surely be an important part of our Anthropocene journey. But so will be assisted adaptation, in which we actively help stressed ecosystems to enable them to adapt sufficiently rapidly to keep pace with the rapidly changing environment. Coral reef scientists are already developing tools we will need to assist reefs to survive, including selective breeding of more heat-tolerant varieties of coral species and methods for coral husbandry that make rearing and transplantation effective for reef restoration.[6] My hope is that, as we build these new tools, we remember that a coral reef is far more than the corals; we must remember the needs of all those other creatures that make a reef what it is. A renewed commitment, and greater effort, to reduce local stressors that affect coral reefs—the overfishing, pollution, physical damage, and general misuse that we have long known are damaging—is also taking place and will surely be helpful in concert with these new approaches.

♦♦♦

Steering our planet safely through Anthropocene seas to places where new kinds of reefs can thrive . . . it sounds so simple, and in chapter 11, I suggested that our acceptance of the need to take on this task was building rapidly. And yet, there is a caveat. Although we now accept the need to act, and while the meme of steering the planet could help convey the enormous importance, and scale, of the task ahead, I do not think we are yet close to accepting the extent of the changes we must make to be successful. Several of Robert Gifford's dragons still stand firmly in front of us.

We remain far more aware of the risks of changing behavior—the financial costs, the cost of changing our routines—than of the risks in continuing as we have been, and we are encouraged to not change our ways by the fact that the global system will not respond quickly to any actions we take. If there is no early evidence that what we are doing to rein in carbon emissions is having real effects on climate, why pay the costs of these efforts? Better still, why not make token efforts, and bask in the glory of having acted, while not making any really big or lasting efforts to change? There remain plenty of reasons to not act or to backslide, to revert to behaving as we have been, and continue to damage the biosphere. Plus there continues to be enthusiastic encouragement to backslide by all those voices from individuals and corporations who stand to lose if the pattern of our economy is altered. We need something more.

We need a radical rethinking of the relationship between humanity and the rest of the biosphere, because this can provide the short-term and immediate benefits to correcting poor environmental practices (fig. 25). This rethinking becomes more important as humanity becomes more urban and less connected to the natural world. We need it because we are not going to become sufficiently motivated to act by the knowledge that acting now

Figure 25. For the person familiar with them, it is possible to feel truly at home on a coral reef. Feeling at home in nature is one step toward recognizing emotionally that we are part of the biosphere. Not separate, certainly not superior, just one part of this amazingly complex, diverse, and multifaceted thing we call life on this planet. Photo © Christina Saenz de Santamaria of One ocean One breath.

reduces risks several decades from now. We need strong present-day reasons for acting to protect the biosphere.

By relationship to the biosphere, I refer to our attitude to the natural world. The conventional Western perspective places humanity next to or above, and certainly outside, the rest of the biosphere instead of inside as an intrinsic part of the biosphere. This tradition of human exceptionalism and an external biosphere *other* is long-standing but not eternal; it has spread with the globalization of market-based economic systems to become the day-to-day perspective of many people whose own cultural traditions may not have maintained such a view. It is one reason that the science of ecology was slow to be sensitive to the effects of humanity on natural systems—the socio- part of socioecological systems.

♦♦♦

I won't embark on an extended discussion of how this worldview developed, or when. Suffice to say that in Europe, between Roman times and the thirteenth century, the evolving structures of English and Continental law gave rights of ownership of land to individuals. Land ownership included ownership of things present on that land except when expressly retained for the use of the crown or the state. By the sixteenth century, notably in the writings of Sir Francis Bacon, Western civilization had taken biblical references to having dominion over nature and molded them into the idea that we humans were separate from, superior to, and entitled to use the rest of creation. At that time, this entitlement meant not too much.[7]

That exceptionalist view still pervades global society and grows stronger as we urbanize and have less direct contact with nature. But now is a time when seeing ourselves as separate, special, and entitled does mean something because we have become powerful enough to make real changes on this planet.

This anthropocentric view of our relation to the environment is best revealed in the consideration of individual property rights. Western law on this subject has been the major influence in the development of international law, as seen in the deliberations of the United Nations, including the 1948 Universal Declaration of Human Rights. This document bluntly affirms in article 17 the right of every individual person to own property, including real property (meaning land).[8]

The 2007 United Nations Declaration on the Rights of Indigenous Peoples, which recognizes clear rights under traditional land tenure systems, struggles to reconcile different perspectives on our relation to the environment in which we live. This document obligates states to provide legal recognition and protection of such property for their Indigenous populations.[9] While doing so, it also acknowledges the existence of systems of land ownership that can be very different to, and not easily conformable with, our Western notions of personal real property, but it accommodates by requiring states to provide the necessary veneer of legality according to the established law of that state. The struggle to reconcile disparate land tenure systems continues, but I fear that, even as we grant rights so long denied to Indigenous peoples, we are losing useful perspectives on our relation to nature. We are Westernizing the legitimate claims of Indigenous peoples to belong to the places they inhabit and hiding (if not erasing) the fundamental difference between their perspectives and the majority Western view.[10]

The European colonization of the Americas, Australia, much of Africa, and parts of Asia, accompanied by a too-healthy sense of racial exceptionalism, created the myth of unowned land that could be claimed by colonizing nations and sold or handed out to settlers. In this process, nature was totally commodified. Modern Western-style nations have built economies in which nature has no rights other than the right to be used by people.[11] In such a societal frame, not only are natural resources there to be used, but there is almost a moral obligation to use them. Within all such societies, it is common to find attitudes conferring respect on owners who manage their land sustainably, but it is also common to find owners who strip profits from their land with little concern for the future. Such behavior is usually fully in compliance with existing law.

With the natural world commodified, and tracts of land owned either by the state or by individuals or groups living within that state and abiding by its laws, ownership rights are commonly treated as including any form of use of that land or its resources. Rights even include patterns of use that rape the land, monetizing whatever value its resources had, and leaving the land degraded, unproductive, and scarcely functioning ecologically.[12]

Legal systems evolve, of course, and common practice tends to lag. In advanced countries, environmental legislation has been developed to protect neighbors and the public from losses incurred through inappropriate use by owners of nearby land. This legislation restricts how waste and by-products of economic activity are to be managed or disposed of, which chemicals, in what quantities, can be dispersed over the environment, what quantities of water can be removed from surface waters or groundwater accessible to owners of adjacent land, which nonnative species or biological agents are permitted to be introduced to a location, and the extent to which local topography can be altered. Such law is administered at national, regional, and municipal levels in instruments as specific as a municipality's by-laws governing the building of a private home on a single building lot. Integrated in this body of law are rules governing the harvest of forest products, minerals, or fisheries from crown or state land or lakes, rivers, and the ocean. Such harvests usually are taken from unowned, open stocks or from resources owned in common, and the governing laws have complexities to deal with equitable sharing of the harvest. This body of broadly environmental law seldom recognizes any rights in nature other than the right to be used sustainably.[13]

I see this worldview as precisely what we should have expected: humans use a commodified nature while being governed by legal systems that create rules to minimize the negative impacts *on other humans* of the uses being made of nature. Natural selection creates humans who act with strong, local, short-term self-interest to improve their lives relative to those of their neighbors. And humans create societies with governing rules that ease the inevitable conflicts among individual members. The result is an economy that serves the present needs of its members. Unfortunately, we happen to have been creative enough to become collectively very powerful, and our actions are altering the world in substantive ways. Our alterations are changing fundamental properties of natural systems such as climate or nutrient cycles, thereby greatly reducing the opportunities for other creatures and for our own offspring.

Three types of action will ease this neglect of the biosphere. The first is to further strengthen the law, forcing us to behave in ways that are environmentally sustainable to the degree necessary to protect our individual, long-term interests, including interests of our descendants, as well as interests of other humans. The second is to extend the law to include specified rights for nonhuman parts of the biosphere. The third is for societies to articulate and adopt a revised worldview that recognizes our individual existence as parts within the biosphere that depend on other parts of it for our own survival and success. These three ways forward are not mutually exclusive. Indeed, I suspect that without at least some shift in worldview, the extent of progress on either of the other two paths will be insufficient to get us out of our current crisis.

♦♦♦

Strengthening the law is a continuous, not necessarily linear, process as we recognize the need to regulate our activities and protect the rights of other people, including those who live in the future. Our collective actions to solve DDT, acid rain, and ozone depletion have all required strengthening of nations' laws. So have the widespread decisions to phase out phosphate use in order to combat eutrophication of lakes and river systems. In my region, there has been a progressive tightening of municipal regulations governing the construction of homes on waterfront property, as understanding of their impacts on aquatic systems has grown.

Extending the law has also been ongoing. When the thirteen American colonies declared their independence from Britain, they wrote about *all men* being created equal, but they meant *all white men with property*. Over time, rights have been extended to less wealthy men, to women, to people of other ethnicities and different gender identities. The pace of change has been uneven and the sequence of changes not necessarily rational; the pattern has been different in different nations. Now there are signs that law is being changed in many nations to give rights to some other animals.

In my own career, I have watched as the study of biology and psychology became limited voluntarily by self-imposed restrictions on what was appropriate management of living vertebrates. The restrictions were imposed and monitored by institutional ethics committees, and by the editorial policies of technical journals that published such research, but this was done in a climate of growing intolerance of "mistreatment" of animals. Legal frameworks continue to tighten.

As a freshman student in introductory zoology I was exposed to the "decerebrate frog preparation"—a frog with the front half of its head removed by making a vertical cut just behind the eyes down to and through the upper jaw. The creature survives this gruesome operation long enough to "perform" certain locomotory and other behaviors during the lab class, a repulsively bizarre way to demonstrate certain aspects of higher and lower brain function in vertebrates. As a graduate student, I prepared such specimens (the frogs are lightly anaesthetized for the cutting), and as a young faculty member in the early 1970s, I witnessed them still being used. I doubt that anyone would dare use this "preparation" in a routine undergraduate teaching lab today—the mass vivisection of dozens of creatures can no longer be justified to make such a simple point.

I have also witnessed gradual extension of concern from higher vertebrates—mammals and birds—to lower vertebrates, and now to octopus and squid, and perhaps to other invertebrates. I learned to grow comfortable with the inanity of having to fill out animal care protocol forms for approval of research in which I would sit in the ocean quietly watching fish freely going about their business. We have not yet reached the point of anaesthetizing corals before sawing them into fragments for mounting on PVC pegs or other platforms to propagate them or use them in physiological studies! Nor have we done much to alter our attitudes to the vivisection of plants. If cabbages could scream, few of us would eat coleslaw, the enthusiasm for vegetarian diets would be reduced, and perhaps we'd talk about vegetable rights.

Experimental study of learning and other aspects of behavior often relies on reward and punishment as a way of helping the organism demonstrate its learning capabilities. Today the severity of punishment is carefully evaluated, and many experiments done by psychologists in the 1960s and 1970s would probably not be approved. Harry Harlow famously reared rhesus monkeys deprived of their mothers and given a choice between a cold wire frame that offered milk and a warm, cuddly frame that did not. He discovered the importance to primates of the mother-infant bond, severely traumatizing numerous macaques for life in the process. Such severe deprivations of a primate wouldn't be permitted today.

Of greater significance, because so many more individual organisms are affected, has been our evolving attitude to the care and eventual slaughter of domestic animals. Advanced countries have extensive law concerning such details as the amount of space per penned individual, the separation of offspring from mothers and littermates, the provision of adequate food, water, and shelter, and especially the treatment of animals being transported and at the abattoir. Yes, we still raise them to kill and eat them, but they are treated more humanely than they used to be, and the legal penalties for failing to comply with the law can be serious. These changes in our treatment of experimental or farm animals, and the legal and other devices that encourage them, are signs of a progressive alteration in our view of the rights of other species. That same alteration has fueled our desire to conserve whales, gorillas, pandas, and polar bears and driven the growing antagonism to keeping animals in zoos or using them in circuses.[14]

♦♦♦

Granting rights to individual animals is not the same as granting rights to nature itself. The slowness with which we have extended rights beyond the higher vertebrates illustrates this difference. Granting rights to nature itself, to a stream, a prairie, or a forest, is a bold step indeed. And yet we are beginning to move in that direction. In his posthumous 1966 essay, "The Land Ethic," Aldo Leopold wrote, "In short, a land ethic changes the role of *Homo sapiens* from conqueror of the land-community to plain member and citizen of it. It implies respect for his fellow-members and also respect for the community as such." Leopold's land ethic defined the gist of what is needed in our expanded worldview, and we have been moving slowly toward that goal ever since.[15]

In 1972, law professor Christopher Stone of the University of Southern California published an article in the *Southern California Law Review*. Entitled "Should Trees Have Standing? Toward Legal Rights for Natural Objects," the essay subsequently morphed into a book in 1974 and has been republished a number of times, most recently as an expanded third edition in 2010. Stone argued that there is no legal impediment to a decision to grant rights to nature, even though to suggest such a step may be surprising.[16]

Speaking for U.S. law, he pointed out that there was a long history of extending rights to a greater and greater range of people, including women, children, ethnic and religious minorities, the cognitively challenged, and individuals in a vegetative state. More tellingly, he pointed to the wide range of inanimate entities now recognized in law—"trusts, corporations, joint ventures, municipalities, Subchapter R partnerships, and nation-states to mention just a few." He noted that ships, routinely referred to as feminine, had long had an independent jural life. United States law has gone even further since 1972. The 2010 ruling by the Supreme Court in *Citizens United v. Federal Election Commission* explicitly extended the constitutional right to free speech to corporations, labor unions, NGOs, and similar "persons," including political action committees. There is no logical barrier to extending rights just as broadly to other nonhuman persons. All that is needed is a legal decision to do so, and several countries have now taken that step.

Setting aside actions taken by individual municipalities or neighborhood groups, of which there are a number, the first nation to grant rights to nature was Ecuador. Its innovative new constitution, adopted in 2008, included language giving legal rights to nature, or *Pachamama,* to "exist, persist, maintain and regenerate its vital cycles, structure, functions and its processes in evolution."[17] Bolivia followed in 2010 with its Law of the Rights of Mother Earth, which drew heavily on Indigenous philosophy in recognizing that Mother Earth or Pachamama is "a dynamic living system comprising an indivisible community of all living systems and living organisms, interrelated, interdependent and complementary, which share a common destiny."[18] Pachamama is recognized as sacred by many Andean societies. This law specifies rights of nature as the right to life, to diversity, to clean air and water, to equilibrium, to restoration, and to living free of pollution.

Subsequent approaches to giving legal rights to nature have been more narrowly focused. New Zealand passed national laws in 2014 and 2017 that respectively designated the Te Urewera wilderness region and the Whanganui River as legal persons. In each case, human guardians are appointed to act on behalf of the natural system. Colombia, in 2016, recognized the Atrato River as an entity with rights, while separately recognizing all animals, and particularly the Andean bear, as subjects with rights. In 2018, Colombia recognized the Colombian Amazon River ecosystem as an entity with rights. India recognized the rights of the Ganges (Ganga) and Yamuna Rivers, and the Gangotri and Yamunotri glaciers, the originating places of the two rivers, in 2017. Similar actions have been taken at state levels in Australia and Mexico and at local levels in other jurisdictions.

Establishing rights in law does not bring immediate compliance, and worldviews do not change overnight. The concept of rights for nature is radical. Strong opposition from interests vested in resource industries has made Ecuador's constitution less effective than many hoped. In 2007, then-president Rafael Correa, who was promoting the new constitution, attempted a bold initiative to protect the Yasuní National Park, a biosphere reserve within the Ecuadorean Amazon that sits atop a billion barrels of oil (the Ishpingo-Tiputini-Tambococha, or ITT, oil field). Correa proposed that nations around the world commit funds to a trust set up for the purpose by the U.N. Development Programme. The trust was to grow to $3.6 billion, representing 50 percent of the revenue that Ecuador would realize if it extracted the oil, and profits from the trust were to support social development in Ecuador. The Yasuní-ITT initiative was intended to keep the oil in the ground and was pitched as a scheme to keep 400 million metric tons of carbon dioxide out of the atmosphere. Perhaps because too many countries saw the scheme as an attempt to extort funds while taking no action (other than not drilling), or perhaps because of the global recession in 2008, the fund did not grow as hoped, and under continuing pressure from creditors and the oil industry, Correa eventually killed the plan in 2013 when only $13 million had been received. The Yasuní reserve has been opened up to oil extraction even though a sizable majority of Ecuadoreans favor its protection. Ecuador's innovative constitution helped push the establishment of the Yasuní-ITT initiative but in the end could not push strongly enough against the tide of resource development. With Correa's departure in 2017, the country is more than ever beholden to the oil industry to sustain its GDP.

However, in a 2016 article in *World Development*, Craig Kauffman of the University of Oregon and Pamela Martin of Coastal Carolina University provided a different perspective.[19] They argued that even though Ecuador's 2008 constitution was too weak to combat the strong push to develop oil reserves in Yasuní, it is playing a role of growing importance in other lower-profile cases, shifting Ecuadorean attitudes toward conservation and away from resource extraction. Progress can be slow, but the consensus among legal scholars is that once jurisdictions recognize rights of nature, attitudes and outcomes shift progressively as cases come before the courts. It is too early to see how legal systems will evolve, but evolution appears to be happening in many places around the globe.

I find it interesting that the push to recognize rights of nature has had only modest success to date among the countries of North America and Europe. There is a small but growing number of municipal, state, and provincial measures in place or under way that achieve this goal, but national laws have not been altered to grant such rights. Admittedly, laws governing environmental damage tend to be strong in these countries, but that is not the same as granting rights to nature. Certainly, these are all countries in which protection of individual property rights is robust; that can conflict with giving rights to nature. Still, I am confident that, globally, something very interesting is happening to legal systems and that as we change our laws, we will inevitably change our worldviews.

♦♦♦

Steering our planet toward a safe, quasi-Holocene future conjures up in my own mind, if not yours, an image of intrepid sailors at the wheel of a glorious vessel fighting to find and hold a course through dangerous seas. The vessel carries our friends and family, including those friends we have made among other entities in the biosphere. I think it's an apt metaphor, apart from missing the fact that the seas are dangerous because our own past misbehavior has made them so. It's a huge task, but a noble one, and one we must tackle.

I have a friend who long ago advanced the argument that we do not have a responsibility to the planet to avoid damaging the biosphere. In his view, other species have come and gone, each seeking to maximize reproductive success regardless of impacts on other species, so why should we be held to a higher standard? Until recently, he worked for a prestigious environmental NGO, and I know his tongue was in his cheek, but he raised a valid point. While coral reefs and other resilient ecosystems can reveal to us the substantial benefits for all that accrue when different species cooperate with one another in intricate ways, there is no scientific argument that says we humans must act ethically to care for the biosphere. If one does not want to use a religious argument, what can we raise as a cogent reason to act this way?

We can plead self-interest, arguing that if we knew the details of how we depend on the rest of the biosphere for our own well-being, we would see the value of acting in ways that sustain that biosphere. But if we do this, there will be those among us who will claim, with some justification, that our science and technology are fully capable, or can become fully capable, of replacing every service the natural world provides for us. We don't need nature, except as decoration.

The hubris in such a view is numbingly expansive! While I can argue that, capable as we are, we are not yet Masters of the Universe and should not gamble with our only home in this way, that argument will fall on deaf ears, and there is little I can say or do to change peoples' opinions. Even the achievements in conquering acid rain or ozone thinning, which I've discussed as evidence that we are capable, can be used to bolster the view that with our science and technology we can do anything we want. While self-interest is real, and a potent additional reason for steering our planet toward a quasi-Holocene future, it is not sufficient. We need a coherent argument that supports responsible, sustainable management of our impacts on the biosphere—an argument for us becoming ethical members of the planet.

Changing our worldview to one that sees us as intrinsically connected to the rest of the biosphere will help. Indeed, I think it is essential. But we need even more than that. We need a nonreligious argument for ethical behavior, and that argument is surprisingly easy to find. To do so, we begin by recognizing those attributes that make us wonderful among living beings. We are not the only species with intelligence, with a capacity to use tools to extend our native abilities, or with the ability to anticipate the future. Nor are we the only species capable of communication beyond the present and the objective. But we have all these capacities in abundance. We can see ten, fifty, even hundreds of years ahead to what the consequences might be of actions we might take. We can communicate those consequences to one another in considerable detail. We can debate logically concerning the better of several alternative paths forward. We have happened rather than chosen to evolve this way, but this is the way we are. Having such abilities, and being as powerful as we now are, we do not have the luxury of refusing to notice the detrimental changes we are causing. We are not just altering the biosphere—we are diminishing its richness, its productivity, and its resilience. We are increasing the entropy of the biosphere.

We are not the same as the chytrid fungus, *Batrachochytrium dendrobatidis*, which is causing massive die-offs and extinctions of frog species around the world; we can appreciate the damage we are doing. Simply because we can understand, appreciate, and inform about this damage, we must strive to do less of it. We do not have the right to do otherwise, because that would condemn our own offspring to live in a severely altered and degraded world while knowing that their own kind caused the loss of wondrousness on this planet. What a legacy to carry!

♦♦♦

The ever-worsening environmental crisis is providing us with plenty of evidence that we do need to change course. The stories making that evidence accessible to people (who do not necessarily put environment front and center every day of their lives) are now being told more effectively than ever before. Many of us, unfortunately, are experiencing some of these stories firsthand, and all of us are just two or three degrees removed from someone who has lived through a savage weather event, witnessed or been part of a forced migration because of environmental collapse, seen a forest cleared or a reef reduced to rubble. It is rapidly becoming impossible to claim, with a straight face, that we do not need to change how humanity tends to its affairs. Or, more to the point, change how we each tend to our own affairs. A logical argument favoring the status quo, in which we continue to overtax the environment in so many ways, can no longer stand up to the evidence of the damage we are causing, and a growing majority of people now recognize and accept that change has got to come. The changes now happening in legal systems, and the shifting perspectives that underly them, are a force that will grow stronger in time, giving us the added motivation to put in the effort needed to correct our misuse of the planet. The only concern I have now is whether we will grab the planetary tiller sufficiently quickly to retain a reasonable portion of the ecological integrity on which our lives depend. Time will tell.

What would happen if nations around the world granted rights to their coral reefs? This is something that could happen if environmental NGOs pushed for such action. Rather than arguing only that the Great Barrier Reef or the Mesoamerican Barrier Reef or that little reef off the coast of a small village in the Philippines should be protected because of the economic value of the goods and services it provides, we could demand that every reef be protected because it has the right to exist and thrive! Such an idea sounds crazy the first time it is spoken, but by repeating this idea, it starts to sound perfectly natural. Why shouldn't a reef have rights at least as strongly protected as those of a little fish on the IUCN Red List or a large corporation?[20] But time is short, and we still may fail to retain coral reefs with anything resembling the functionality they now have, let alone the wondrousness. Their disappearance, if it does occur, will be a difficult lesson but one we will all need to remember:

> Yes, there was a time back in the twenty-first century, a time before humanity relearned the humility necessary to recognize that we are a part of the biosphere. A time before we saw that we must act in ways that fulfilled our own needs, while protecting the capacity of the biosphere to continue to support all other life as well. During that time, our ancestors did great damage to the ecological integrity of our shared planet, and one of the casualties was coral reefs, wondrously majestic cornucopias of life, the most diverse and productive of all marine systems, and a pinnacle of evolutionary exuberance. We cannot bring them back. We suffer because of the loss of what they could provide. And we remember them and celebrate them, so that the next time some part of the biosphere cries out in distress, we will heed, we will act, and we will strive to steer wisely, thereby continuing this incredible journey we travel together on this strangely wondrous little planet, Earth.

ACKNOWLEDGMENTS

Conceived in 2016 and slowly gestating through 2017, this book was an outline with chapter headings by March 2018. Casual discussions over that time surely guided its structure and content, but the details of such are lost in the mists of time. By late 2018, I had a first rough draft. Several individuals read all or parts of this draft and were not afraid to tell me where it was wanting. Jon Lovett Doust, a plant ecologist, was the first serious reader, and I thank him particularly for his insights and his encouragement. Marine scientists Drew Harvell, Jake Kritzer, Pete Mumby, John Ogden, and Bob Steneck all provided cogent comments and useful suggestions. Rain forest ecologist Meg Lowman and freshwater ecologist Norman D. Yan helped identify the parts that would be obscure to someone who does not already know reefs. Finally, my good friend and total nonscientist, Allan McLeod, read the full text and told me in no uncertain terms where it descended so deeply into the weeds that nobody but a scientist would care. I hope the final version is substantially better than what each of them read.

For illustrations, apart from maps, I reached out to a number of skilled photographers—both long-term colleagues and people I had never met. Without exception, these talented people were supportive. Luiz Rocha, whose images I had long admired on Twitter, provided figures 10, 12, 17, 23, and one part of 16. Andy Hooten, a wonderful colleague over twenty years, provided figures 9, 20, and two parts of figure 3, as well as helpful advice on converting images to monochrome. Kris Bruland, a photographer whom I know only from his flickr page, https://www.flickr.com/photos/133511324@N05/albums/, was generous with images, providing figure 13, part of figure 14, and two parts of figure 16. Several years ago, I saw an amazing photo by Eusebio and Christina Saenz de Santamaria, of One ocean One breath, www.oneoceanonebreath.com. They provided figures 11 and 25. Donald L. Mykles, an invertebrate physiologist at Colorado State University, readily provided figure 2 when I contacted him out of the blue—someone else I have never met. Ove Hoegh-Guldberg, whom I've known since he was an undergraduate with a broken arm, provided one part of figure 3. Longtime friend Bob Steneck provided another part of figure 3. Alex Vail, now a wildlife photographer, whose thesis research featured in chapter 5, provided two parts of figure 14. Isabelle Côté generously offered images when approached, providing one part of figure 16. Suzan Meldonian, an underwater photographer based in Boca Raton, Florida, https://www.niteflightphoto.com/, opened my eyes to the wonders of black water photography and provided figure 18. Patrick L. Colin, Coral Reef Research Foundation, Koror, Palau, http://coralreefpalau.org, a colleague I was last in contact with in the early 1980s, responded promptly when I reached out and provided figure 21. Another longtime colleague, J. E. N. (Charlie) Veron, quickly sent me the Adobe Illustrator file for figure 22, despite being in the middle of moving house at the time. Bryson Hoe and Kapua Roback of Polynesian Voyaging Society and 'Ōiwi Television Network, Hawai'i, willingly provided figure 24 (the photo was taken by Kaipo Ki'aha) when I contacted the society and asked. Finally, Victor Huertas, a graduate student completing his Ph.D. at James Cook University, and another person I have never met, generously provided several photos, one of which we selected as the cover image for the book.

Alina Szmant and Mary Alice Coffroth provided needed technical help with coral and symbiont physiology. Geoff Jones, Peter Doherty, Ove Hoegh-Guldberg, Lexa Grutter, Mark Hixon, and Terry Hughes brought me up to date or filled in forgotten details of reef ecology. Randy Olson improved my understanding of storytelling, Ben Victor and Lexa Grutter helped chase images, and Steve Mussman, Alex Brylske, and L. Patrick Boyer gave marketing advice. To all these people, and some I have certainly forgotten, my sincere thanks.

At Yale University Press, acquisitions editor Jean Thomson Black and her assistant Elizabeth Sylvia have always been responsive and supportive, while Jean's deft editorial hand steered me in appropriate directions. As I write, Jean has passed me along to production editor Joyce Ippolito, who has assured me she will shepherd the book through the rest of the way until its emergence as a bound physical item. Her first step was to put me in the delightfully capable hands of manuscript editor Laura Jones Dooley, who has done a wonderful job of diminishing my tendency to use three words when one will suffice. These four women have been, for me, the face of a professional yet human institution. Yale University Press treats its authors well. Thank you, Jean, for seeing the possibilities in my manuscript.

Last, my wife, Donna, has endured my endless hours at the computer, assuming it will all be worthwhile in the end. But she is the same person who put up with all those days and weeks of absence while I was in the field on a coral reef somewhere far away. I think she knew when we married that our life might be like that. I've always enjoyed her support and know I am permanently in her debt.

NOTES

1 Beginnings

1. Some male crabs wave their claws as a social signal to attract mates and ward off other males, but usually they wave only one; this gesture is clearly distinct from the "two claws high and wide apart" defensive display. Fiddler crabs, common on intertidal mudflats through the tropics, are particularly conspicuous when they use this one-claw signal: the waved claw is usually much larger than the other and is brightly colored.
2. T. G. Wolcott, and D. L. Wolcott, "Larval Loss and Spawning Behavior in the Land Crab *Gecarcinus lateralis* (Fréminville)," *Journal of Crustacean Biology* 2 (1982): 477–485.

2 Serendipity in Deep Time

1. Rotenone is a natural plant product discovered by Indigenous South Americans and used by them to collect fish for food. It affects the capacity of the gills to extract oxygen from the water, causing the fish to asphyxiate. It breaks down rapidly in water, and eating fish killed by it is not harmful. Rotenone was at one time widely used for removing undesirable species of fish from water bodies.
2. During his first circumnavigation, while sailing up the coast of Queensland, Cook ran aground on a reef a few minutes before 11:00 p.m. on June 11, 1770, yet he'd had 17 fathoms (30 meters) of water beneath the keel moments before the ship struck. He was too busy holding his ship together to reflect that day, or for several days later, but on August 16, having repaired his ship and set out once more, he experienced a second perilous time when the ship was up current of a reef, the current was

freshening, and the wind had died. His journal entry contained this description: "All the dangers we had escaped were little in comparison of being thrown upon this reef, where the Ship must be dashed to pieces in a Moment. A reef such as one speaks of here is Scarcely known in Europe. It is a Wall of Coral Rock rising almost perpendicular out of the unfathomable Ocean, always overflown at high Water generally 7 or 8 feet, and dry in places at Low Water. The Large Waves of the Vast Ocean meeting with so sudden a resistance makes a most Terrible Surf, breaking Mountains high, especially as in our case, when the General Trade Wind blows directly upon it." Entry for August 16, 1770, after having been almost swept up onto the outer barrier, from *The Journals of Captain James Cook on His Voyages of Discovery*, vol. 1: *The Voyage of the Endeavour, 1768–1771*, ed. J. C. Beaglehole (Cambridge: Hakluyt Society at the University Press, 1955).

3. A primary goal of the science of taxonomy is to group organisms in ways that reflect their evolutionary origins and relationships. Thus, all creatures possessing nematocysts are grouped in the phylum Cnidaria because it is quite improbable that something as complex and unique as a nematocyst would have evolved more than once. Typically, the organisms within a phylum are grouped into classes, containing orders, which in turn contain families. Each family includes one or more genera, and each genus contains one or more species. The species is the fundamental unit—a group of individual organisms that are usually capable of reproducing with each other. This hierarchy of phylum, class, order, family, genus, and species can be made more complex by creating sub- and super-groupings—subphyla, superorders—to better display the patterns of relatedness in a particular group of creatures. The hierarchical structure is subject to change as scientific understanding of a group of organisms grows, but it provides a robust framework enabling biologists to comprehend the relationships among all creatures, in much the same way that the chemists' periodic table does for all elements.
4. The prevailing form in the jellyfishes and box jellies is the free-swimming medusa, a bell- or umbrellalike organism with numerous tentacles trailing from its outer edge and a single, downwardly directed mouth in the center under the bell. With a bit of topological gymnastics, the medusa and the polyp can be visualized as structurally similar, one sitting mouth up and the other floating mouth down.
5. Deepwater corals do not have symbiotic algae living within their tissues and are far less capable of rapid calcification than shallow-water species. Reef-building corals all possess symbiotic algae.
6. I've gone into these taxonomic details because it's important to understand that the word *coral* defines a number of different organisms, some closely and some distantly related to each other. The ability to build a calcified skeleton is widely distributed, but not universal, among cnidarians.
7. Sponges and several now-extinct creatures were the primary contributors to these

ancient reefs, but corals played a part. El Capitan, an 85-meter-high sheer cliff and the most prominent peak in the Guadalupe Mountains of West Texas and New Mexico, is part of the Capitan Reef that extended almost 600 kilometers along the border of the shallow Delaware Basin about 290 million years ago during the Permian. The even more impressive Devonian Reef (350 million years old) is visible as a 350-kilometer-long barrier formation up to 50 kilometers wide along the northern boundary of the Canning Basin in Western Australia. Anyone who has hiked the 700-kilometer Bruce Trail along Ontario's Niagara Escarpment from Niagara Falls to the tip of the Bruce Peninsula, or continued farther north along the western shore of Manitoulin Island, has walked perhaps a third the length of an even older, immense Silurian reef, built by tabulate corals about 420 million years ago.

8. Judith Lang, "Interspecific Aggression by Scleractinian Corals: 2. Why the Race Is Not Only to the Swift," *Bulletin of Marine Science* 23 (1973): 260–279.
9. Tiles fastened to the substratum provide a new site for settlement of larvae of many marine creatures such as corals and are useful in ecological studies of such organisms.
10. D. Ross Robertson, "Fish Feces as Fish Food on a Pacific Coral Reef," *Marine Ecology Progress Series* 7 (1982): 253–265; David R. Bellwood, Andrew S. Hoey, and J. Howard Choat, "Limited Functional Redundancy in High Diversity Systems: Resilience and Ecosystem Function on Coral Reefs," *Ecology Letters* 6 (2003): 281–285.
11. Darwin's observations of depth limits were made very indirectly by casting a lead (a 7-to-14-pound lead weight armed with tallow to pick up evidence of the nature of the substratum) tied to a graduated line. Based on his own observations and a survey of reports by others, he stated, "We may, I think, conclude that in ordinary cases, reef-building polypifers do not flourish at greater depths than between twenty and thirty fathoms [36–54 meters' depth]." Charles Darwin, *On the Structure and Distribution of Coral Reefs; also Geological Observations on the Volcanic Islands and Parts of South America . . .*, Minerva Library of Famous Books, edited by G. T. Bettany (London: Ward, Lock, 1890), 68.

Darwin's subsistence hypothesis was an elegant thought experiment that simultaneously explained the development of fringing reefs, barrier reefs, and atolls as due to the same set of processes. Corals grew in shallow waters at the shore of a landmass. If that landmass subsided over time and the corals continued upward growth, the shoreline would retreat, leaving the initially fringing reef separated from the shore as a barrier reef. If the landmass was an island and subsided completely beneath the waves, and the corals continued growing up toward sea level, the barrier reef would be left as a ring of reef surrounding a shallow lagoon, in otherwise deep water.

12. K. R. Ludwig et al., "Strontium-Isotope Stratigraphy of Enewetak Atoll," *Geology* 16 (1988): 173–177; Terrence M. Quinn and Arthur H. Saller, "Geology of Anewetak Atoll, Republic of the Marshall Islands," in *Geology and Hydrogeology of Carbonate Islands*, Developments in Sedimentology 54, edited by H. Leonard Vacher and Terrence M. Quinn (Amsterdam: Elsevier, 2004), 637–666.
13. Their less equatorial location may also play a role. Reef growth is less rapid in cooler waters.
14. The geological history of Bermuda is complex, driven by the dynamics of the Pleistocene. During the Pleistocene there were at least twenty episodes of cooling, with massive formation of glaciers in northern latitudes and greatly lowered sea level. These were interspersed with warmer interglacial periods when sea levels could be higher than today. Over the 2.6 million years of the Pleistocene, these fluctuations in sea level and temperature resulted in a sequence of episodes of coral reef growth followed by episodes of reef erosion, producing quantities of sand that were piled into impressive dunes. What exists today is the summation of all the reef building, erosion, and dune building that took place in those past times.
15. For this account, I draw on David Hopley, Scott G. Smithers, and Kevin E. Parnell, *The Geomorphology of the Great Barrier Reef: Development, Diversity and Change* (Cambridge: Cambridge University Press, 2007); J. M. Pandolfi and R. Kelly, "The Great Barrier Reef in Time and Space: Geology and Paleobiology," in *The Great Barrier Reef: Biology, Environment and Management*, edited by Pat Hutchings, Mike Kingsford, and Ove Hoegh-Guldberg (Collingwood, Vic.: CSIRO, 2007), 17–27; Jody M. Webster and Peter J. Davies, "Coral Variation in Two Deep Drill Cores: Significance for the Pleistocene Development of the Great Barrier Reef," *Sedimentary Geology* 159 (2003): 61–80; J. F. Marshall and P. J. Davies, "Internal Structure and Holocene Evolution of One Tree Reef, Southern Great Barrier Reef," *Coral Reefs* 1 (1982): 21–28; and Peter J. Davies, "Evolution of the Great Barrier Reef—Reductionist Dream or Expansionist Vision," *Proceedings, 6th International Coral Reef Symposium*, vol. 1 (Townsville, Qld.: 6th ICRS Executive Committee, 1988), 9–17.
16. Reefs on these plateaus were surely the initial sources of corals and other reef creatures when the continental shelf finally became suitable for reef growth.

3 Ever Wonderful, Always Different

1. How I got to Australia was a journey fully as serendipitous as how I got to Hawai'i several years before. I knew that as a foreign student I had to leave the United States once I completed my studies. I knew I wanted to continue working on coral reefs. And I surmised that the chance of working on coral reefs from Canada was remote. I needed to find a place where I could work on coral reefs for a couple of years and

then hightail it back to the States, hopefully to Hawai'i, to live happily ever after. Knowing little about the reality of the international community to which successful academics belong, and discussing my predicament with nobody who knew more than I did, I decided that Australia was the only place I could go, work on reefs, and still have a chance to get back into the U.S. academic community. So, I wrote letters to heads of biology departments in six Australian universities telling them about myself and asking if they were hiring. Miraculously, University of Sydney took this bait, and I became a faculty member there, remaining for twenty years doing research on the Great Barrier Reef. Again, not a process for advancement I'd recommend to any aspiring graduate student today.

2. In the early 1980s, as president and vice president respectively of the Great Barrier Reef Committee (which earlier had created the Heron Island Research Station), Hal Heatwole and I were instrumental in transforming a venerable but moribund organization into the Australian Coral Reef Society, a far more dynamic science organization that continues today. (Even coral reef scientists can sometimes achieve great things!) Incidentally, the ACRS rightly bills itself as "the oldest organization in the world concerned with the study and protection of coral reefs." Australian Coral Reef Society, https://australiancoralreefsociety.org.
3. Looe Key was designated a national marine sanctuary in 1981; John Pennekamp State Park off Key Largo had been designated in 1960. Both are now included within the Florida Keys National Marine Sanctuary, established in 1990.
4. One Tree Reef is zoned as a scientific zone within the Great Barrier Reef Marine Park and is accessible only with a science permit. One Tree Island and its reef have been used for scientific research since 1965. The three small huts have grown into a modest field research facility, now administered by the University of Sydney, https://sydney.edu.au/science/our-research/research-facilities/one-tree-island.html. Much of the early history and a description of the site are captured in a book: Harold Heatwole, *A Coral Island: The Story of One Tree Island and Its Reef* (Sydney: Collins, 1981).
5. The most powerful form of ecological research remains the use of simple field experiments to test explicit hypotheses. To do such an experiment, it is necessary to find a set of individuals—whether individual coral heads, fish defending territories, patch reefs, spurs, grooves, and so on—to serve as the replicate experimental units. On coral reefs, finding replicates can be a challenge.
6. This lagoon is 100+ kilometers wide, 30–60 meters deep, and filled with reefs. It lies between the outermost reefs on the edge of the continental shelf and the Queensland coast—very different to the tiny lagoons of One Tree or Heron Reefs.
7. The Fourteenth ICRS, originally scheduled for July 5–10, 2020, was postponed until 2021 because of the coronavirus pandemic. As I write, it is scheduled for Bremen, Germany, July 18–23, 2021.

8. Unfortunately, most of us were male in those days, but "guys" has never been gender-specific when referring to reef scientists.
9. Robertson documented the way in which social interactions between a male and his harem of females prevent mature females from changing sex and becoming male. In a classic field experiment, he removed several males from the reef and then watched the females over the next several days. Within hours, the largest of the females in a harem was performing characteristically male behaviors; within two days "she" was spawning as a male; and within two weeks, "she," now definitely "he," was producing viable sperm. D. R. Robertson, "Social Control of Sex Reversal in a Coral-Reef Fish," *Science* 177 (1972): 1007–1009.
10. That funding would be available for such a venture reveals just how threatened coral reefs are by our activities. It would be easier, as was the practice in earlier times, simply to bury the reefs with the new construction.

4 So Many Ways of Being

1. M. H. A. Keenleyside, "The Behaviour of *Abudefduf zonatus* (Pisces, Pomacentridae) at Heron Island, Great Barrier Reef," *Animal Behaviour* 20 (1972): 763–775.
2. Although species can be defined on the basis of breeding compatibility—if they do, they are the same species—we have no such tidy rules to delineate genera, orders, classes, or phyla. The goal of taxonomists is to group organisms that are related—have arisen from a common ancestor—but individual taxonomists vary in how finely they divide what is really a continuum. Tradition, convention, and common sense keep taxonomy from being chaotic, but there are often rearrangements of species among these higher levels as particular groups become better known or new technologies provide additional insights concerning relationships. That's one of the reasons students today learn about five kingdoms of organisms whereas I learned about only plants and animals.
3. The naming of species of animal is controlled by a series of rules that (1) give emphasis to priority and (2) seek to group evolutionarily related species together. A wide variety of characteristics—morphological, genetic, molecular, physiological, even behavioral—are used to make decisions concerning evolutionary relatedness, and the first name given in describing a new species or genus is usually the one that is used. When the New Guinea and Guam descriptions of this drab little fish were recognized as being of the same species, the earlier (Guam) name, *biocellatus*, replaced *zonatus*. When the *Abudefduf* hodge-podge was split apart, the Heron Island species reverted briefly to Cuvier's original 1830 genus *Glyphisodon*, before becoming *Chrysiptera* when *Glyphisodon* itself was split apart (in this case, the species name was altered to a feminine spelling to match the feminine *Chrysiptera*, from the masculine form matching *Glyphisodon*—Latin scholars would have

understood this). While these rules may seem complex and more linguistic than scientific, they do help create order in the way scientists name creatures.

4. Peter F. Sale, "Recruitment, Loss, and Coexistence in a Guild of Territorial Coral Reef Fishes," *Oecologia* 42 (1979): 159–177.
5. We had submitted specimens of many taxa to taxonomists in museums; some may now have been formally named. Our article: P. F. Sale, P. S. McWilliam, and D. T. Anderson, "Composition of the Near-Reef Zooplankton at Heron Reef, Great Barrier Reef," *Marine Biology* 34 (1976): 59–66.
6. J. Frederick Grassle, "Variety in Coral Reef Communities," in *Biology and Geology of Coral Reefs*, Vol. 2: *Biology* 1, edited by O.A. Jones and R. Endean (New York: Academic, 1973), 247–270.
7. One of Hutchinson's most widely read papers was titled "The Paradox of the Plankton." It explored the relatively high species richness of the plankton of freshwater lakes. He speculated on the question of how so many species could be evolutionarily successful while doing rather similar things in the same lake—the phytoplankton in particular are mostly single-celled plants that all float about in the upper water layers, using photosynthesis to capture the energy in sunlight and use it to build organic models from simpler inorganic compounds in the water column.
8. Bill Gosline taught me that a scientist's worth is in the ideas he or she produces, not in the amount of grant money his or her research brings to the university. I used that knowledge in often unsuccessful conversations with deans and vice presidents concerning relative merit of faculty in departments I chaired much later in my career. Too often academic administrators fail to see beyond the dollars.
9. Actually, my conversation was with Jon. Many years later he became Joan in revealing his trans identity. Roughgarden went on to write a remarkable book on sexuality and gender across the animal kingdom: Joan Roughgarden, *Evolution's Rainbow: Diversity, Gender, and Sexuality in Nature and People* (Berkeley: University of California Press, 2004).
10. Times have changed. What was true in the 1970s and 1980s is less true now. Tropical universities such as James Cook University in Townsville, Queensland, have first-rate ecologists on their faculties, and cutting-edge ecology is now being done by individuals who live in the tropical regions they study.

5 Exuberant Richness

1. Although I emphasize fishes here, a similar story can be told for each component of the coral reef fauna. For regional-scale patterns of distribution and for local-scale patterns of habitat use, for demographic patterns, for social structures and reproductive modes, we could tell similar stories about crustaceans, mollusks, or the corals themselves.

2. Mark W. Westneat and Peter C. Wainwright, “Feeding Mechanism of *Epibulus insidiator* (Labridae; Teleostei): Evolution of a Novel Functional System,” *Journal of Morphology* 202 (1989): 129–150.
3. C. Seabird McKeon, and Jenna M. Moore, “Species and Size Diversity in Protective Services Offered by Coral Guard-Crabs,” *PeerJ* 2 (2014): e574, https://doi.org/10.7717/peerj.574.
4. The slender, often translucent pearlfishes, Carapidae, make homes within the anal cavities (technically the cloaca rather than the anus, but who’s quibbling) of sea cucumbers and starfishes, and the hosts do whatever they can to try and dislodge them. I suppose these could be seen as social relationships, although trying to evict an anal intruder seems a rather unlikely start to a long-term social bond.
5. I. Karplus, R. Szlep, and M. Tsurnamal, “Goby-Shrimp Partner Specificity. I. Distribution in the Northern Red Sea and Partner Specificity,” *Journal of Experimental Marine Biology and Ecology* 51 (1981): 1–19.
6. I regret throwing all these Latin names at you, but just as many reef species remain undescribed, many others have never gained common names. We recognize snapping shrimps, for example, but that name covers hundreds of similar-looking species of *Alpheus*. I refuse to take the easy way out, translate the Latin name, and then pretend that is a commonly used name for that creature!
7. Although territoriality is well studied in many coral reef damselfishes, Ross Robertson’s research on Indian Ocean surgeonfishes revealed that several species are routinely territorial, protecting algal food resources. D. Ross Robertson, Nicholas V. C. Polunin, and Kimberley Leighton, “The Behavioral Ecology of Three Indian Ocean Surgeonfishes (*Acanthurus lineatus*, *A. leucosternon*, and *Zebrasoma scopas*): Their Feeding Strategies, and Social and Mating Systems,” *Environmental Biology of Fishes* 4 (1979): 125–170; D. Ross Robertson and Steven D. Gaines, “Interference Competition Structures Habitat Use in a Local Assemblage of Coral Reef Surgeonfishes,” *Ecology* 67 (1986): 1372–1383. But Robertson also found considerable complexity, with only some members of the population being territorial and some individuals switching between territorial and nonterritorial behavior over weeks. This suggests that there is lots more to learn about social organization in this family of reef fishes. Robertson was also the first to do a detailed study of how schooling permits a parrotfish (the Caribbean striped parrotfish, *Scarus iseri*) to swamp the defense of damselfish territories, but he also drew attention to the presence of some territorial individuals within the parrotfish population. D. R. Robertson et al., “Schooling as a Mechanism for Circumventing the Territoriality of Competitors,” *Ecology* 57 (1976): 1208–1220.
8. R. E. Thresher, “Field Experiments on the Problem of Species Recognition by the Threespot Damselfish, *Eupomacentrus planifrons* (Pisces: Pomacentridae),” *Animal Behaviour* 24 (1976): 562–569.

9. A good summary of this research is in Peter F. Sale, “Recruitment, Loss and Co-existence in a Guild of Territorial Reef Fishes,” *Oecologia* 42 (1979): 159–177.
10. I reported briefly on this behavior in Peter F. Sale, “Reef Fishes and Other Vertebrates: A Comparison of Social Structures,” in *Contrasts in Behavior: Adaptations in the Aquatic and Terrestrial Environments,* edited by Ernst S. Reese and Frederick J. Lighter (New York: John Wiley and Sons, 1978), 313–346.
11. Steve Strand, “Following Behavior: Interspecific Foraging Associations among Gulf of California Reef Fishes,” *Copeia* (1988): 351–357.
12. Hans Fricke first mentioned joint hunting by morays and groupers in *The Coral Seas* (London: Thames and Hudson, 1972), but Ilan Karplus appears to have been the first to detail the practice. Ilan Karplus, “A Feeding Association between the Grouper *Epinephelus fasciatus* and the Moray Eel *Gymnothorax griseus,*” *Copeia* (1978): 164.
13. Redouan Bshary et al., “Interspecific Communicative and Coordinated Hunting between Groupers and Giant Moray Eels in the Red Sea,” *PLoS Biology* 4 (2006): e431, doi: 10.1371/journal.pbio.0040431.
14. Alexander L. Vail, Andrea Manica, and Redouan Bshary, “Referential Gestures in Fish Collaborative Hunting,” *Nature Communications* 4 (2013): 1765, doi: 10.1038/ncomms2781; Alexander L. Vail, Andrea Manica, and Redouan Bshary, “Fish Choose Appropriately When and with Whom to Collaborate,” *Current Biology* 24 (2014): R791–R793.
15. I once climbed that hill and walked about on top, but the day was cloudy and I could not even see the outer barrier, let alone a path through it. Still, my pacing about on top of the hill ensured that I had walked in James Cook’s footsteps. Alex Vail undoubtedly climbed that same hill; more relevant for our story, he got to know Redouan Bshary during Bshary’s visits to Lizard Island and ended up as a student at Cambridge, doing field research back home on Lizard! The Lizard Island Research Station is at https://australianmuseum.net.au/get-involved/amri/lirs/.
16. Conrad Limbaugh, “Cleaning Symbiosis,” *Scientific American* 205 (1961): 42–49.
17. The ectoparasites are a diverse group including many creatures that can survive in the reef environment when not on a fish. Indeed, larval gnathiid isopods hop on and off, spending only hours to days on a host fish and living nonparasitic benthic lives as adults. Meanwhile, adults of the cymothoid isopod *Anilocra* sp., which use their legs to hook themselves firmly into the face of their host and grow large enough to obstruct the host’s vision, stay far longer, lose the ability to swim, and are likely not capable of reattaching to a fish if they become dislodged. Although gnathiids and *Anilocra* have direct development, other common ectoparasites, such as digenean and aspidogastrean trematodes (flukes), have an intermediate (usually mollusk) host occupied by the juvenile stage(s).
18. In the Caribbean several small, closely related gobies, *Elacatinus* spp., come closest

to filling the role of *Labroides* species in the Indo-West Pacific in that they clean throughout their lives. However, these cleaners are seldom more than 2 centimeters in length. They feed also on coral mucus, and they appear to have a much smaller impact on the lives of their coral reef neighbors than do *Labroides* in the Pacific. I've never seen anything in the Caribbean resembling the busy cleaning stations manned by two or three *Labroides* that are so conspicuous on Pacific reefs.

19. Initial experiments were reported by Marsh Youngbluth and by George Losey, both at University of Hawai'i: Marsh J. Youngbluth, "Aspects of the Ecology and Ethology of the Cleaning Fish, *Labroides phthirophagus* Randall," *Zeitschrift für Tierpsychologie* 25 (1968): 915–932; and George S. Losey Jr., "The Ecological Importance of Cleaning Symbiosis," *Copeia* (1972): 820–833. More recent experiments are reported in Peter A. Waldie et al., "Long-Term Effects of the Cleaner Fish *Labroides dimidiatus* on Coral Reef Fish Communities," *PLoS ONE* 6 (2011): e21201, doi:10.1371/journal.pone.0021201; and A. S. Grutter et al., "Parasite Infestation Increases on Coral Reefs without Cleaner Fish," *Coral Reefs* 37 (2018): 15–24.
20. Examples are Karen M. Cheney, and Isabelle M. Côté, "Mutualism or Parasitism? The Variable Outcome of Cleaning Symbioses," *Biology Letters* 1 (2005): 162–165; and Redouan Bshary, Simon Gingins, and Alexander L. Vail, "Social Cognition in Fishes," *Trends in Cognitive Sciences* 18 (2014): 465–471.
21. Ana Pinto et al., "Cleaner Wrasses *Labroides dimidiatus* Are More Cooperative in the Presence of an Audience," *Current Biology* 21 (2011): 1140–1144; S. Tebbich, R. Bshary, and A. S. Grutter, "Cleaner Fish *Labroides dimidiatus* Recognise Familiar Clients," *Animal Cognition* 5 (2002): 139–145.
22. I. M. Côté, C. Arnal, and J. D. Reynolds, "Variation in Posing Behaviour among Fish Species Visiting Cleaning Stations," *Journal of Fish Biology* 53, supp. A (1998): 256–266.

6 Wonderful Surprises

1. This assumption that nature is in balance also underlies (or underlay, because the world is different now) our approach to management of national parks—minimize deleterious human impacts and the ecosystem will go on in perpetuity just as it is right now, unchanging, in balance, at equilibrium.
2. An important factor in how the pelagic life phase disrupts the inheritance of success from one generation to another has to do with how big a role chance plays in determining survival when the overwhelming majority of larval fishes will die. When one in a thousand larvae survive on average, doubling the rate of survival (or doubling the number of larvae produced) still means that nearly every larval fish fails to survive—small increases in fitness of adults convey little benefit on offspring in the competition to survive and get back to the reef. The work on territorial

damselfishes was published in a series of papers between 1972 and 1982. The most widely cited was Peter F. Sale, "Maintenance of High Diversity in Coral Reef Fish Communities," *American Naturalist* 111 (1977): 337–359. The most comprehensive was Peter F. Sale, "Recruitment, Loss, and Coexistence in a Guild of Territorial Coral Reef Fishes," *Oecologia* 42 (1979): 159–177.

3. Peter F. Sale, "Stock-Recruit Relationships and Regional Coexistence in a Lottery Competitive System: A Simulation Study," *American Naturalist* 120 (1982): 139–159.
4. In a closed environment, individual competitive success translates into enhanced reproductive success in a more direct way than in an open system. In such a world, a slight competitive advantage translates into a rapidly increasing proportion of all larvae being larvae produced by the superior species. In such a world, even if individual places on a reef are captured randomly by the first larva that happens to swim by, the greater abundance of larvae of the competitively superior species ensures that it takes over all sites in just a few generations, eliminating any other competitor species.
5. Stephen P. Hubbell, "Tree Dispersion, Abundance, and Diversity in a Tropical Dry Forest," *Science* 203 (1979): 1299–1309.
6. Steve Hubbell is also author of an influential, and decidedly nonequilibrial, book, Stephen P. Hubbell, *The Unified Neutral Theory of Biodiversity and Biogeography* (Princeton, N.J.: Princeton University Press, 2001).
7. See Peter F. Sale, and Rand Dybdahl, "Determinants of Community Structure for Coral Reef Fishes in an Experimental Habitat," *Ecology* 56 (1975): 1343–1355; and Peter F. Sale, Jeffrey A. Guy, and Warren J. Steel, "Ecological Structure of Assemblages of Coral Reef Fishes on Isolated Patch Reefs," *Oecologia* 98 (1994): 83–99.
8. One lesson this study taught me was that some individual fishes, even though scarcely 10 centimeters in length, could live a long time. While there was plenty of change from census to census among my coral patches, many fish were almost certainly the same individuals seen on the same patch over years. Censusing the coral patches again and again gave me the joy of visiting old friends! I vividly remember the little *Pomacentrus amboinensis* with a peculiar dark blue star on her forehead (a color defect allowing the pigmented *dura mater* surrounding her brain to shine through, and one I've not seen before or since), who lived at one of the smaller coral patches. That fish was a young adult when I first saw her, perhaps 18 months old. Twenty censuses and 10 years later, she was an old lady about 10 centimeters in length. I hope she sent many clutches of larvae off to sea over her life, and I wonder how long she lived after my visits ended. I miss seeing her, although I'm sure she never felt so strongly about me.
9. When a scientific paradigm is well established and widely accepted within the research community, as niche theory was in the 1970s, it is very difficult to dislodge it. Even now, in a world that is palpably not in a stable equilibrium state, when ecolo-

gists can tick off the underlying assumptions that limit niche theory—stationarity, resources fully used, closed system—many of us still implicitly assume that niche theory defines the usual case for any community of organisms unless there are data to the contrary.

10. One feature I seem to have been unable to convince many of my colleagues about concerns how the results I obtained using relatively small coral patches should influence how we interpret our observations of larger patches or stretches of contiguous reef. At this larger scale, there will obviously be many more fish and more species present, and many researchers have shown proportionally less pronounced change in numbers or in species composition through time than I reported for small patches. The usual interpretation is that the larger-scale studies reveal a consistency through time that is compatible with ideas from conventional niche theory. I interpret those larger-scale results instead as due to two things. First, there are the consequence of averaging over space—the small-scale variability is still there, but averaging evens out the results apparent at the larger scale. Second, sampling at the larger scale cannot provide as precise a count as sampling at the smaller scale—with less precision, there is less evidence of variability through time.
11. Looking back, it is uncanny how broadly accepted was the view that coral reef habitat was usually filled with species. Those of us who studied fish all saw how they rapidly colonized artificial reefs or natural reefs that had had their resident fishes removed. And damselfishes were not the only ones who fought over space. Not until we carefully counted numbers of small fish on small coral patches did we see that space was filled variably.
12. This study was published as Peter J. Doherty, "Tropical Territorial Damselfishes: Is Density Limited by Aggression or Recruitment?" *Ecology* 64 (1983): 176–190.
13. It takes real skill to remove 10-centimeter-long fish from a reef using a tiny spear powered by a rubber band held between the fingers. And it's uncanny how little fishes know, the moment you get in the water, that you plan on spearing instead of just watching them.
14. I've always believed that pilot experiments should be modest in scale. Peter, not so much. In another memorable effort during his Ph.D. program, Peter wanted to collect data on breeding and egg production by *P. wardi.* A logical way forward was to offer nest sites that could be more easily inspected than the places the males usually chose. They did often nest under dead *Tridacna* shells (*Tridacna maxima* is a common 15-to-25-centimeter-long giant clam), and so Peter decided that *Tridacna* shells cemented to flat bases so that they would stand upright would be good nests. Without asking the fish, Peter embarked on a massive *Tridacna* shell scavenger hunt, followed by an artisanal program of curio production. When he was finished, the shore of One Tree Island held what seemed like an army of clam shells standing proudly upright and all being washed in the surf. But once he deployed these nests

out on the reef, Peter discovered that *P. wardi* was not interested in using a *Tridacna* shell sticking up like a Texaco sign! None of them ever got used, but, hey, science is like that—bright people bustling about, running off in all sorts of useless directions, and every now and then discovering something important.

7 How Nemo Found His Way Home

1. J. H. Choat and L. M. Axe, "Growth and Longevity in Acanthurid Fishes: An Analysis of Otolith Increments," *Marine Ecology Progress Series* 134 (1996): 15–26.
2. The Reynolds number, R_e (named after physicist Osbourne Reynolds, who did pioneering work on fluid dynamics in the late 1800s), is the ratio between inertial and viscous forces that governs movement in a fluid. The Reynolds number, R_e, is calculated as: $R_e = U\, l/v$ in which U is swimming speed in centimeters per second, l is total body length in millimeters, and v is the kinematic viscosity of sea water. (I've given you that equation, just to remind you that this is science. No need to remember it!) When viscous forces dominate (R_e<300), swimming is slow and relatively inefficient, but when inertial forces dominate (R_e>1000), swimming is efficient and relatively more rapid.
3. Jeffrey M. Leis, "The Pelagic Stage of Reef Fishes: The Larval Biology of Coral Reef Fishes," in *The Ecology of Fishes on Coral Reefs*, edited by Peter F. Sale (San Diego, Calif.: Academic Press, 1991), 183–230.
4. In the mid-1950s Jack Randall had used a light suspended over the stern of a boat to attract juvenile manini during his Ph.D. studies and observed that they certainly seemed to be *actively* swimming into the shallows rather than being carried there on the tide. Many other fish biologists had used lights in this way, but the idea of a light trap, floating in the ocean, luring in larval fishes, was something new.
5. For damselfishes and cardinalfishes, the 13.5 centimeter-per-second current in the flume represents about nine to twelve body lengths per second, nine to twelve times as fast, relatively, as Michael Phelps on a good day. At that speed, they managed some two hundred to five hundred laps of an Olympic pool—not exactly wimpy!
6. I am relying primarily on three articles for information on orientation of larvae: I. C. Stobutzki and D. R. Bellwood, "Nocturnal Orientation to Reefs by Late Pelagic Stage Coral Reef Fishes," *Coral Reefs* 17 (1998): 103–110; Jeffrey M. Leis and Mark I. McCormick, "The Biology, Behavior, and Ecology of the Pelagic, Larval Stage of Coral Reef Fishes," in *Coral Reef Fishes: Dynamics and Diversity in a Complex Ecosystem*, edited by Peter F. Sale (San Diego, Calif.: Academic Press, 2002), 171–199; and Jeffrey M. Leis, "Are Larvae of Demersal Fishes Plankton or Nekton?" *Advances in Marine Biology* 51 (2006): 57–141.
7. Gabriele Gerlach et al., "Smelling Home Can Prevent Dispersal of Reef Fish Larvae," *PNAS* 104 (2007): 858–863.

8. Technically, otoliths float within the labyrinthine organ, comprising the sacculus, the utriculus, and the three semicircular canals. Their movements within these fluid-filled chambers play an important role in balance, orientation, and hearing by the fish.
9. I am glossing over the work involved in determining that it was possible to bathe eggs in tetracycline and then see a fluorescent mark in the otolith that grew weeks later; first they had to master the aquaculture of *Pomacentrus amboinensis*!
10. Geoff Jones exhibits the same workaholic traits as Peter Doherty—maybe it's a New Zealander characteristic. In any event, they both impress and amaze me. The study is in G. P. Jones et al., "Self-Recruitment in a Coral Reef Fish Population," *Nature* 402 (1999): 802–804.
11. Again, I am glossing over the many trials, failures, and fruitless hours it took to reach this point.
12. Such things happen when you are a faculty member who does field research in amazing places and to whom students flock from all over. Information about Mahonia na Dari: https://mahonianadari.org/; and Walindi Plantation Resort: http://www.walindifebrina.com/.
13. The results were not perfectly congruent—a few tagged recruits did not have parents among the genotyped adults, and some recruits with parents at the island did not show any signs of tetracycline. This discrepancy suggests that tetracycline tagging is not 100 percent effective under field conditions and that some replacement of breeding pairs occurred during the season.
14. This is yet another unanswered question probably worth investigating. It's not unusual to find a place where a certain species is very common, even though it is rare or absent everywhere else over kilometers of reef. How did the members of the group find each other after larval life, and why are they found at only one place? Coral reefs throw up mysteries like this continuously.
15. I'm glossing over some complexities here: it was the ratio of ^{137}Ba to other Ba isotopes that was measured, rather than the concentration of ^{137}Ba isotope, and the reason both 9 of 15 collected clownfish recruits and 8 of 77 collected butterflyfish recruits lead to estimates of 60 percent returning home is simply because the tagging program included all breeding clownfishes but only 17 percent of the breeding butterflyfishes. Glenn R. Almany et al., "Local Replenishment of Coral Reef Fish Populations in a Marine Reserve," *Science* 316 (2007): 742–744.
16. There used to be two primary science journals: *Science* and *Nature*. But in recent years, *Nature* has revealed unexpected fecundity, spinning off a whole stable of high-quality offspring. *Nature* journals now number about forty, including the grand dame herself, which has published since 1869. By contrast, *Science*, publishing since 1880, has spun off just five offspring in recent years. Not sure why I felt compelled to tell you this.

17. Glenn R. Almany et al., "Larval Fish Dispersal in a Coral-Reef Seascape," *Nature Ecology and Evolution* 1 (2017): 0148, doi:10.1038/s41559-017-0148.
18. A metapopulation is a group of local populations connected to one another by some level of exchange (via larvae in this case) of individuals. There is presently considerable theoretical interest among conservation biologists in defining the spatial scale of metapopulations because this information would enhance the effectiveness of small marine protected areas by ensuring that such sites are at appropriate distances from one another to facilitate some exchange among them.
19. While the work at Kimbe Bay continued, Geoff, Simon, and Serge applied the barium and parentage techniques to establish the larval dispersal patterns for the coral trout (*Plectropomus leopardus*), a much larger, commercially, and recreationally important species of grouper on the Great Barrier Reef. They found evidence of larvae dispersing 200 kilometers from the site of spawning to the site of settlement, but also of self-recruitment to small no-take fishery reserves and export of other larvae from fishery reserves to fished areas within the region sampled. This vindicates those who have argued that networks of protected areas can sustain fishery species. And it may finally quiet the fishery scientists who look down on fish ecologists who study so-called silly little fishes. Sometimes we discover important things.
20. John E. Randall, "A Contribution to the Biology of the Convict Surgeonfish of the Hawaiian Islands, *Acanthurus triostegus sandvicensis*," *Pacific Science* 15 (1961): 215–272.
21. This is a reasonable question precisely because we do know a lot about reproductive and parental behavior in damselfishes. *Acanthochromis* is so far unique. Ross Robertson documented its reproductive behavior in D. R. Robertson, "Field Observations on the Reproductive Behaviour of a Pomacentrid Fish, *Acanthochromis polyacanthus*," *Zeitschrift für Tierpsychologie* 32 (1973): 319–324. All other species in this family, so far as we know, care for their eggs but not their hatched young.
22. Only somebody as rabidly crazy as scientist turned filmmaker, and my friend, Randy Olson, would actually attempt to follow tunicate larvae around, but Randy earned a Ph.D. from Harvard by doing so at Lizard Island in the early 1980s. Read about it in Richard Randolph Olson, "The Consequences of Short-Distance Larval Dispersal in a Sessile Marine Invertebrate," *Ecology* 66 (1985): 30–39.

8 Wondrous Reef Chemistry

1. O. Hoegh-Guldberg et al., "Coral Reefs under Rapid Climate Change and Ocean Acidification," *Science* 318 (2007): 1737–1742. This has become one of the most heavily cited papers on coral reefs of all time. I remain proud to be a coauthor, even if I still don't really understand aragonite saturation curves.

2. Quoted in James Bowen and Margarita Bowen, *The Great Barrier Reef: History, Science, Heritage* (Cambridge: Cambridge University Press, 2002).
3. I'm using the words *algae* and *plant* in a nontechnical sense meaning "organisms that photosynthesize." The algae are a diverse group of creatures only some of which are closely related to the plants in our gardens. Until recently, these symbiotic algae had been placed in the kingdom Protista (all single-celled organisms with chromosomes contained in a nucleus), and true plants in kingdom Plantae, but over the past decade or so molecular genetic data have forced a substantial reorganization of the higher classifications of life. At present, what used to be called the Eukaryota (all organisms other than bacteria and viruses) has been divided into more than twenty phyla (singular phylum), which are placed into supergroups, each containing one or more phyla. In this scheme, zooxanthellae, other dinoflagellates, a number of other single-celled forms, and the brown algae are grouped as Chromalveolates; the green plants and green and red algae are the Archaeplastida or Plantae; and the animals, fungi, and some single-celled organisms are grouped as Opisthokonts. Still other groups comprise the Amoebazoa and the Excavates. Personally, I still live in a plants, animals, protists, and bacteria world, and I seem to be doing fine even if, by doing so, I am not fully acknowledging the amazing diversity of life on Earth.
4. The symbiotic dinoflagellate, living within coral cells, is a rounded cell packed with chloroplasts and does not possess flagella; these are evident only during its free-living phase.
5. This yearlong science expedition to Low Isles, central Great Barrier Reef, led by C. Maurice (later Sir Maurice) Yonge in 1928–1929 and funded by the Royal Society of London, marks the world's first major coordinated effort to understand the biology of coral reef systems. Yonge was an invertebrate physiologist and experimentalist rather than a taxonomist, the more typical biological expedition leader of those times. The British team had limited tropical experience, so they certainly brought fresh eyes to a largely unknown ecosystem. The results of the year at Low Isles were published in seven volumes between 1930 and 1968 and contributed significantly to our understanding of coral reef ecosystems.
6. While I've always thought the idea of using Pacific atolls as sites for atomic testing was unsavory (especially because the populations of these atolls were forcibly moved elsewhere to make the testing possible), I recognize that a lot of ecological science on coral reefs got done as part of atomic testing programs.
7. The flow cytometry technique places the biological specimen into a closed chamber filled with water, with one inlet, one outlet, and a constant flow through the chamber. This makes possible an indirect measurement of metabolic activities by monitoring the change in dissolved gases such as oxygen, or nutrients such as phosphate, between the inflow and the outflow ports. Sargent and Austin treated

the shallow bench as if it were a closed chamber containing the bench reef community.

8. Their results were published in 1949 in a technical report, and subsequently as Marston C. Sargent and Thomas S. Austin, "Biologic Economy of Coral Reefs," *Geological Survey Professional Paper* 260-E (1954): 293–300.
9. These calculations express the production or consumption of oxygen in terms of amount for each 1-centimeter-wide strip of reef bench; the bench was 300 meters across from seaward to lagoonal side, so a 1-centimeter-wide strip represents 3 square meters of reef environment.
10. More influential because it was published by better-known authors and in a higher profile journal but also because it looked more deeply into the topic of productivity on reefs.
11. The AEC was spending considerable sums on environmental research at that time, all directed to understanding the impact of atomic weapons and/or atomic power plants on natural systems. Howard Odum subsequently led a major study of rainforest ecology in which a portion of Puerto Rico's El Yunque rain forest was irradiated.
12. 10.4 Mt = explosive force equivalent to 10.4 million metric tons of TNT.
13. During my first year in Hawai'i, I almost went to Enewetak as one of two graduate students hired as summer lab managers. At the last minute the offer was canceled when administrators realized I was not a citizen. In those days only male U.S. citizens could visit Enewetak. Apparently, it was thought that a noncitizen might be more likely than others to tell everyone that rockets from California were being landed in the lagoon! And as for females? Well, really. . . . Ah, the cold war. Restrictions eased in later years, and then the research station closed down.
14. Howard T. Odum and Eugene P. Odum, "Trophic Structure and Productivity of a Windward Coral Reef Community on Eniwetok Atoll," *Ecological Monographs* 25 (1955): 291–320.
15. Evergreen broadleaf forest production data are from Jianyong Ma et al., "Gross Primary Production of Global Forest Ecosystems Has Been Overestimated," *Scientific Reports* 5 (2015): 10820, doi: 10.1038/srep10820.
16. Bruce Gordon Hatcher, "Coral Reef Primary Productivity: A Beggar's Banquet," *Trends in Ecology and Evolution* 3 (1988): 106–111.
17. Blue-green algae, Cyanophyta, or Cyanobacteria are organisms with cells lacking a nucleus. They are a division of the domain Eubacteria. They photosynthesize but are only very distantly related to the green, red, and brown algae, three related groups of photosynthesizing organisms that all possess nucleate cells (within domain Eukaryota along with green plants, fungi, and us).
18. For information on these algal symbionts and their interactions with corals, I have benefited from advice from Mary Alice Coffroth and Alina Szmant, and the

following articles: Iliana B. Baums, Meghann K. Devlin-Durante, and Todd C. LaJeunesse, "New Insights into the Dynamics between Reef Corals and Their Associated Dinoflagellate Endosymbionts from Population Genetic Studies," *Molecular Ecology* 23 (2014): 4203–4215; Daniel J. Thornhill et al., "Host-Specialist Lineages Dominate the Adaptive Radiation of Reef Coral Endosymbionts," *Evolution* 68 (2013): 352–367; D. A. Poland and M. A. Coffroth, "Trans-Generational Specificity within a Cnidarian-Algal Symbiosis," *Coral Reefs* 36 (2017): 119–129; and Todd C. LaJeunesse et al., "Systematic Revision of Symbiodiniaceae Highlights the Antiquity and Diversity of Coral Endosymbionts," *Current Biology* 28 (2018): 2570–2580.

19. I find many things remarkable about the intricacy of the relationships among the coral and its symbionts. I also find it amazing, given the importance of the host-symbiont connection, that most corals rely on picking up suitable algal symbionts from the environment early in juvenile life. (Certain species, in which larvae are brooded within the gastric cavity, gain some symbionts before being released.) It should have been simple in all corals to evolve a way to transmit symbionts from parent to offspring within the egg. That this has not happened confirms that, as usual, evolution builds complexity that is good enough to allow creatures to survive and reproduce; no fine engineering is involved. Incidentally, in being intracellular, these algal symbionts are more like chloroplasts and mitochondria than like most other commensal or parasitic organisms, which live among a host's cells rather than within them. It is now generally accepted that chloroplasts and mitochondria evolved from what were originally symbiotic bacterial cells.
20. One difficulty is that these are primarily asexually reproducing organisms. This makes the usual species definition based on sexual reproduction difficult. Molecular analyses have revealed amounts of genetic variation between clades that are as great or greater than those between genera or families of conventionally reproducing organisms.
21. Altogether, nine clades have now been recognized of which the most important in corals are clades A and B, both common in the Caribbean but occurring elsewhere as well; clade C, globally widespread, and nearly universal on Australasian reefs; and clade D, found mostly in the Indian Ocean, Red Sea, and Western Pacific. Each clade contains several types or species. *Symbiodinium* taxonomy is a puzzle that is only now being sorted out using genetics and molecular biology. I've relied on LaJeunesse et al., "Systematic Revision of Symbiodiniaceae."
22. I rely for the information on symbiont distribution on Andrew C. Baker, "Flexibility and Specificity in Coral-Algal Symbiosis: Diversity, Ecology, and Biogeography," *Annual Review of Ecology and Systematics* 34 (2003): 661–689; Madelaine J. H. van Oppen et al., "Bleaching Resistance and the Role of Algal Symbionts," in *Coral Bleaching: Patterns, Processes, Causes and Consequences*, edited by Madelaine J. H.

van Oppen and Janice M. Lough, Ecological Studies 205 (Berlin: Springer, 2009), 83–102; Poland and Coffroth, "Trans-Generational Specificity within a Cnidarian-Algal Symbiosis"; and LaJeunesse et al., "Systematic Revision of Symbiodiniaceae."

23. In describing the communication and cooperation among bacteria, algal symbionts, and the coral, I don't mean to imply conscious decision-making by any of these creatures. Each performs in particular ways, sometimes in direct response to actions by others, often in response to local chemistry, and the result is a cooperative achievement of complex tasks. In earlier centuries we'd have talked about how these intricate relationships reveal the beauty of God's creation, and many now would talk about the power of evolution. I'm content to talk about the wondrousness that is revealed, content to marvel in the unexpected complexity of the interactions taking place.
24. This two-step dance is more easily displayed as three simple equations: (1) $CaCO_3 \leftrightarrow Ca^{++} + CO_3^{=}$; (2) $CO_2 + H_2O \rightarrow HCO_3^{-} + H^{+}$; and (3) $H^{+} + CO_3^{=} \rightarrow HCO_3^{-}$.
25. CO_2 dissolved in surface waters acts to consume $CO_3^{=}$ in producing HCO_3^{-}, thereby driving the equilibrium between solid calcium carbonate and the dissolved calcium and carbonate ions toward the ions. Furthermore, the more CO_2 in the atmosphere, the more CO_2 dissolves in the water, dissociates, increases the amount of HCO_3^{-} ions at the expense of $CO_3^{=}$ ions, encourages the dissolution of $CaCO_3$ into its constituent ions, and lowers pH. This lowering of pH, called ocean acidification, is a separate, troubling effect on the ocean caused by our CO_2 emissions. Among other things, it enhances dissolution of existing reef and may slow calcification by corals.
26. Two and a half times more rapid in one recent study; 3 times is typical, although there are reports up to 126 times faster. Data are in Aurélie Moya et al., "Study of Calcification during a Daily Cycle of the Coral *Stylophora pistillata*: Implications for 'Light-Enhanced Calcification,'" *Journal of Experimental Biology* 209 (2006): 3413–3419.
27. Bleaching is a physiological stress response; the intimate symbiosis between the coral and its dinoflagellates breaks down, the algae are expelled, and the coral appears pale or translucent. Bleached corals are physiologically challenged and likely to die if the warm conditions continue. Warming-induced bleaching is becoming the primary cause of reef degradation as seas warm.
28. I am not implying that corals and their symbionts consciously do these things, but these things get done and the sequence of events is essential for calcification to happen—not conscious but still wondrous.
29. Data from Alina M. Szmant, and in Tim Wijgerde et al., "Coral Calcification under Daily Oxygen Saturation and pH Dynamics Reveals the Important Role of Oxygen," *Biology Open* 3 (2014): 489–493.
30. A perspective that includes the joy of being in a reef environment, visiting old

friends such as the little *Pomacentrus amboinensis* with the peculiar blue star on her forehead, mentioned in chapter 6. I doubt I could have survived a career as a biologist whose research kept him in the lab, with the white coat and the glossy instruments.

9 What Good Is a Coral Reef?

1. Many will claim that fitness equals value, but fitness, in its evolutionary sense, simply recognizes that some among a set of coexisting but slightly different phenotypes will have attributes that favor that phenotype in its current environment. Such favored phenotypes will be likely to leave more offspring than less favored ones, and those offspring will inherit some or all of the attributes that conveyed favor. Only if environmental conditions persist will they be similarly favored. There is no intrinsic value or sustained value here.
2. The Brundtland Report, https://sustainabledevelopment.un.org/content/documents/5987our-common-future.pdf, also known as *Our Common Future,* was published in 1987 as the final and primary product of the World Commission on Environment and Development, or Brundtland Commission, established as an independent entity in 1984 by the U.N. General Assembly and chaired by Gro Harlem Brundtland, prime minister of Norway. The Rio Declaration on Environment and Development, https://www.cbd.int/doc/ref/rio-declaration.shtml, was the primary agreement of the U.N. Conference on Environment and Development, held in Rio de Janeiro, Brazil, June 3–14, 1992. The Rio Declaration included twenty-seven principles intended to guide sustainable development by the 170 signatory countries, including no. 15, the precautionary principle, and no. 16, the polluter pays principle.
3. Millennium Ecosystem Assessment, *Ecosystems and Human Well-Being: Synthesis* (Washington, D.C.: Island Press, 2005). The assessment was carried out between 2001 and 2005 to assess the consequences of ecosystem change for human well-being and to establish the scientific basis for actions needed to enhance the conservation and sustainable use of ecosystems and their contributions to human well-being.
4. Nonrenewable natural resources such as oil that have not yet been exploited are also often valued, just not in the context of ecosystem valuation. Indeed, some people argue that a nation has an obligation to extract and use its natural resources, building prosperity in the process—sometimes this is expressed as a moral imperative!
5. Entry to this debate about valuation and nature is available via Joshua Farley, "Ecosystem Services: The Economics Debate," *Ecosystem Services* 1 (2012): 40–49.
6. Robert Costanza et al., "The Value of the World's Ecosystem Services and Natural Capital," *Nature* 387 (1997): 253–260.
7. The Economics of Ecosystems and Biodiversity, TEEB, is a global initiative hosted by the U.N. Environment Programme. It produced a set of five linked documents in 2010, available at http://www.teebweb.org/our-publications/teeb-study-reports/.

Together they provide methods and a global evaluation of ecosystem value. These methods have been adopted in much of the subsequent research in this field.

8. Rudolf de Groot et al., "Global Estimates of the Value of Ecosystems and Their Services in Monetary Units," *Ecosystem Services* 1 (2012): 50–61. Robert Costanza et al., "Changes in the Global Value of Ecosystem Services," *Global Environmental Change* 26 (2014): 152–158.
9. Samia Sarkis, Pieter J. H. van Beukering, and Emily McKenzie, eds., *Total Economic Value of Bermuda's Coral Reefs: Valuation of Ecosystem Services,* technical report (Department of Conservation Services, Government of Bermuda), available at https://environment.bm/coral-reef-economic-valuation.
10. Deloitte Access Economics, *At What Price? The Economic, Social and Icon Value of the Great Barrier Reef,* report commissioned by the Great Barrier Reef Foundation (Brisbane: Deloitte Access Economics, 2017), available at https://www2.deloitte.com/content/dam/Deloitte/au/Documents/Economics/deloitte-au-economics-great-barrier-reef-230617.pdf.
11. Access Economics, *Economic Contribution of the GBRMP, 2006–2007,* Research Publication 98 (Great Barrier Reef Marine Park Authority, 2009), available at http://dspace-prod.gbrmpa.gov.au/jspui/bitstream/11017/436/1/Economic-contribution-of-the-GBRMP-2006-2007.pdf. This report provides a comparative table of Great Barrier Reef values back to 2004.
12. Willingness-to-pay approaches use surveys in which participants are asked such questions as, "How much would you be willing to pay per year to ensure that the Great Barrier Reef will continue to offer opportunities for reef vacations?" This approach provides an opportunity value, which is distinct from, and added to, the value in dollars paid out to take reef vacations during a year. While I abhor the idea of valuing the Great Barrier Reef asset or capital using a thirty-three-year life and depreciation of 3.7 percent, this is a standard approach to valuing other kinds of assets, and adopting it probably gains credence across the business community.
13. Deloitte Access Economics, *At What Price?*
14. B. Kushner et al., *Coastal Capital: Jamaica; Coral Reefs, Beach Erosion and Impacts to Tourism in Jamaica* (Washington, D.C.: World Resources Institute, 2011), available online at http://www.wri.org/coastal-capital. Using data from Negril, Ochos Rios, and Montego Bay, Kushner and colleagues determined that ten further years of beach erosion at current rates will reduce annual tourism revenue to Jamaica by $19 million but that continued reef degradation will increase that erosion and the resulting loss in revenue will be $33 million per year.
15. Palm Jebel Ali, due to be completed in two years when I visited in 2003, remains in early 2020 a set of sandy islets slowly being washed away. Palm Jumeirah, by contrast, is a busy urban development of hotels, theme parks, restaurants, and scores of higher-end apartments and homes, most of which stand empty much of the time, unused possessions in unvisited places. In the words of one recent review: "I'll be

honest: I'm not usually impressed by things made big and extravagant for the sake of it. But, there's something impossible to deny about the hubris behind the Palm Jumeirah and, when you see it in person, it sticks with you, for better or for worse." Hubris? No doubt about that, and I've avoided mention of the World, yet another extravagance of artificial islands on the Dubai coast. Harrison Jacobs, "I Stayed at a Hotel on Dubai's Massive Artificial Island Shaped Like a Palm Tree and It's More Surreal Than Any Photos Can Show," *Business Insider*, December 3, 2018, https://www.businessinsider.com/dubai-palm-jumeirah-artificial-island-2018-12.

16. The state of Queensland has jurisdiction over many aspects of use of all land within the park, including islands far offshore, while the federal government has jurisdiction over most of the coastal ocean, so park management is achieved by sets of mirroring legislation in Queensland and federally. The park is managed by an Authority with federal and Queensland representation, with much of day-to-day management done by the state of Queensland with federal funding to do so.
17. GBRMPA is the usual term when referring to the Great Barrier Reef Marine Park Authority.
18. Nonmarket value of ecosystems is often also nonsubstitutable. A coral reef is a wondrous place that enriches all our lives simply by existing. That value cannot be replaced by any stock market portfolio.

10 Why Don't We Seem to Care about Coral Reefs?

1. Excerpts from this report, *Sources, Abundance, and Fate of Gaseous Atmospheric Pollutants*, are at https://www.smokeandfumes.org/documents/16, a compendium of documents revealing how early, and the extent to which, the energy industry was aware of climate effects of CO_2 emissions. The 400 ppm milestone was not passed until 2014, when April readings at Mauna Loa averaged 401 ppm. Readings averaged 315 ppm in March 1958, when instruments were first deployed on the mountain. Data are at https://www.esrl.noaa.gov/gmd/ccgg/trends/data.html.
2. The 2015 report by *InsideClimateNews* is at https://insideclimatenews.org/content/Exxon-The-Road-Not-Taken, and the *Los Angeles Times* story begins at http://graphics.latimes.com/exxon-arctic/#about. The story was also well covered in Naomi Oreskes and Erik Conway, *Merchants of Doubt: How a Handful of Scientists Obscured the Truth on Issues from Tobacco Smoke to Global Warming* (New York: Bloomsbury, 2010). Through membership in the American Legislative Exchange Council (ALEC, since 1973 an organization of mostly right-of-center legislators, corporate leaders, and lobbyists that drafts model bills on selected topics, thereby influencing state and federal legislation) and the Global Climate Coalition (an industry-led, international lobbying group advocating against climate science and active from 1989 to 2001), Exxon has been able to delay or defuse legislative efforts

on climate change. Through more direct lobbying efforts, it has been influential in determining U.S. representation on international bodies such as the Intergovernmental Panel on Climate Change and documents and agreements arising in such bodies.

3. *InsideClimateNews* has a copy of this memo at http://insideclimatenews.org/sites/default/files/documents/1982%20Exxon%20Primer%20on%20CO2%20Greenhouse%20Effect.pdf.
4. The Panama bleaching is described in Peter W. Glynn, "Widespread Coral Mortality and the 1982/83 El Niño Warming Event," *Environmental Conservation* 11 (1984): 133–146. More localized bleaching caused by low salinity or warm or cold water had been seen occasionally in the early 1980s, but nothing stretching over tens of kilometers before this. The Great Barrier Reef has lost more than 50 percent of average coral cover since 1985 (mean of 13.8 percent in 2012 versus 28.0 percent in 1985 on monitored reefs), although climate change is not the only or even the most important cause. Glenn De'ath et al., "The 27-Year Decline of Coral Cover on the Great Barrier Reef and Its Causes," *PNAS* 109 (2012): 17995–17999. See also D. R. Bellwood et al., "Confronting the Coral Reef Crisis," *Nature* 429 (2004): 827–833; and Terry P. Hughes et al., "Global Warming Transforms Coral Reef Assemblages," *Nature* 556 (2018): 492–496. For the Caribbean, see J. B. C. Jackson et al., *Status and Trends of Caribbean Coral Reefs: 1970–2012* (Gland, Switzerland: Global Coral Reef Monitoring Network, IUCN, 2014).
5. The 1992 Rio Declaration on Environment and Development included twenty-seven principles intended to guide sustainable development by the 170 signatory countries, and IPBES released *Summary for Policymakers of the IPBES Assessment Report on Land Degradation and Restoration* for its thematic assessment of land degradation and restoration in March 2018. It can be downloaded at https://www.ipbes.net/event/ipbes-6-plenary.
6. Accessible summaries of what we are doing to our planet are in Johan Rockström and Mattias Klum, with Peter Miller, *Big World, Small Planet: Abundance within Planetary Boundaries* (New Haven: Yale University Press, 2015); and Will Steffen et al., "The Trajectory of the Anthropocene: The Great Acceleration," *Anthropocene Review* 2 (2015): 81–98.
7. Ferdinand K. J. Oberle, Curt D. Storlazzi, and Till J. J. Hanebuth, "What a Drag: Quantifying the Global Impact of Chronic Bottom Trawling on Continental Shelf Sediment," *Journal of Marine Systems* 159 (2016): 109–119; Robert J. Diaz and Rutger Rosenberg, "Spreading Dead Zones and Consequences for Marine Ecosystems," *Science* 321 (2008): 926–929; Denise Breitburg et al., "Declining Oxygen in the Global Ocean and Coastal Waters," *Science* 359 (2018): eaam7240, doi: 10.1126/science.aam7240; L. Lebreton et al., "Evidence That the Great Pacific Garbage Patch Is Rapidly Accumulating Plastic," *Scientific Reports* 8 (2018): 4666,

doi:10.1038/s41598-018-22939-w; L. Jewett and A. Romanou, "Ocean Acidification and Other Ocean Changes," in *Climate Science Special Report: Fourth National Climate Assessment*, vol. I, edited by D. J. Wuebbles et al. (Washington, D.C.: U.S. Global Change Research Program, 2018), 364–392.

8. Although we have been notably far less effective in correcting local damage than we might!
9. This depressing tale is documented in more detail in Peter F. Sale, *Our Dying Planet* (Berkeley: University of California Press, 2011), and in numerous other books, articles, and media reports, including O. Hoegh-Guldberg et al., "Coral Reefs under Rapid Climate Change and Ocean Acidification," *Science* 318 (2007): 1737–1742; J. E. N. Veron, *A Reef in Time: The Great Barrier Reef from Beginning to End* (Cambridge, Mass.: Belknap Press of Harvard University Press, 2008); and *Chasing Coral*, a 2017 Exposure Labs documentary, directed by Jeff Orlowski and available on Netflix or at https://www.chasingcoral.com/. In October 2018, the Intergovernmental Panel on Climate Change (IPCC) released the draft *Summary for Policy Makers* of *Global Warming of* 1.5°C, a comparison of the likely consequences of warming 1.5°C versus 2.0°C, available at https://www.ipcc.ch/sr15/. It specified that "coral reefs, for example, are projected to decline by a further 70–90% at 1.5°C (high confidence) with larger losses (>99%) at 2°C (very high confidence)." Since the world is currently heading toward 3°C or 4°C by 2100, it's clear that the IPCC considers the likelihood of coral reefs surviving anywhere on this planet to be very remote indeed.
10. The Holocene commenced 11,500 years ago as the world warmed following the final Pleistocene glaciation. It has been climatically a particularly stable period, and human civilizations developed within it. It remains possible that if we curtail our greenhouse gas emissions soon and quickly enough, we may avert true runaway melting and retain some glacial ice in polar regions. There will come a time, however, when nothing we can do will stop the melting until all glacial ice is gone. I'm not claiming that the new world we are entering will be one in which our species cannot survive. I am claiming that it will be one that our current agriculture and other technologies are not adapted to, and unable to transition toward quickly, without disrupting the provision of food and other goods and services needed to support our enormous population. The global economy is not a sleek and efficient system able to adapt quickly and seamlessly to disruption in its environment.
11. O. Hoegh-Guldberg et al., "Impacts of 1.5°C Global Warming on Natural and Human Systems," in *Global Warming of* 1.5°C, edited by V. Masson-Delmotte et al. (IPCC, 2018), available at https://www.ipcc.ch/sr15/.
12. Coal miners used to take canaries into the mines with them. Far more sensitive to carbon monoxide than were miners, the canary provided an early warning of

potentially serious problems. When the canary fell off its perch, it was time to get out of the mine. In Britain, coal miners carried canaries until 1986.

13. I remember an early 1980s conversation at Discovery Bay, Jamaica, with an incredulous social scientist who was attempting to build awareness in the local fishing community of the need to conserve some adult fishes. The trap fishermen he worked with knew that juvenile fish came into the reefs from the open ocean as minute larvae, but they did not understand that breeding by the adult fishes on the reefs produced those larvae in the first place. They saw little reason to not take every fish off the reef—more would come from the ocean—and comments by a sociologist from Canada were not changing their minds.
14. While scientists struggle to attend to our poor oral or written communications, it's abundantly clear that effective presentations and papers, ones that are structured as stories, garner more attention and are better remembered. Good storytelling skills pay off in science, even if we scientists don't pay much attention to the need to learn how to do this.
15. "Scientists" is in quotation marks because many of the individuals who feed denialist campaigns on environmental issues have dubious or irrelevant expertise.
16. Terry P. Hughes et al., "Global Warming Transforms Coral Reef Assemblages," *Nature* 556 (2018): 492–496.
17. Embargoing until the publication date is the usual approach. It gives journalists a heads-up and time to write articles that will appear as soon as the article is published. "Global Warming Is Transforming the Great Barrier Reef," *Science Daily*, April 18, 2018, https://www.sciencedaily.com/releases/2018/04/180418141504.htm; Robinson Meyer, "Since 2016, Half of All Coral in the Great Barrier Reef Has Died," *Atlantic*, April 18, 2018, https://www.theatlantic.com/science/archive/2018/04/since-2016-half-the-coral-in-the-great-barrier-reef-has-perished/558302/.
18. Peter Hannam, "'Cooked': Study Finds Great Barrier Reef Transformed by Mass Bleaching," *Sydney Morning Herald*, April 18, 2018, https://www.smh.com.au/environment/climate-change/cooked-study-finds-great-barrier-reef-transformed-by-mass-bleaching-20180418-p4za9m.html.
19. News Corp generally and the *Australian* particularly are widely recognized as a significant outlet for climate change denialism. The *Australian* article: Graham Lloyd, "Not All Scientists Agree on Cause of Great Barrier Reef Damage," April 19, 2018, https://www.theaustralian.com.au/news/health-science/not-all-scientists-agree-on-cause-of-great-barrier-reef-damage/news-story/1960ad7d2010e9f2194e32889ed99260.
20. These media reports are at Kate Wheeling, "Corals Can Withstand Another Century of Climate Change," *Pacific Standard*, April 19, 2018, https://psmag.com/environment/corals-can-withstand-another-century-of-climate-change; and at Josh

Gabbatiss, "First Genetically Engineered Coral Created to Help Save Reefs from Climate Change," *Independent*, April 28, 2018, https://www.independent.co.uk/environment/coral-reef-genetically-engineered-climate-change-great-barrier-global-warming-a8318756.html.

21. Damien Caves, "Australia Pledges Millions of Dollars in Bid to Rescue Great Barrier Reef," *New York Times*, April 29, 2018, https://www.nytimes.com/2018/04/29/world/australia/great-barrier-reef-plan.html; Peter Hannam, "Great Barrier Reef's Five Near-Death Experiences Revealed in New Paper," *Sydney Morning Herald*, May 28, 2018, https://www.smh.com.au/environment/climate-change/great-barrier-reef-s-five-near-death-experiences-revealed-in-new-paper-20180528-p4zhwb.html; Bill McKibben, "How Justin Trudeau and Jerry Brown Can Help Save the Great Barrier Reef," *New Yorker*, May 30, 2018, https://www.newyorker.com/news/daily-comment/how-justin-trudeau-and-jerry-brown-can-help-save-the-great-barrier-reef; "World's Largest Coral Reef Farm Set for Fujairah," *Gulf Today*, June 1, 2018, https://www.pressreader.com/bahrain/gulf-today/20180601/page/2/textview; Ben Smee, "Coral Decline in Great Barrier Reef 'Unprecedented,'" *Guardian*, June 5, 2018, https://www.theguardian.com/environment/2018/jun/05/coral-decline-in-great-barrier-reef-unprecedented. One irony not mentioned in the *New York Times* article—the funds were given not to the strong Australian reef science community or to the management agency responsible for the reef but to the Great Barrier Reef Foundation, a body without the capacity to do research or management, one led by people with ties to the oil industry, and one that did not even request the funds.
22. A survey of news on social media would yield a similarly chaotic view, but with scarcely any detail so that readers in the dark would never be brought into the light.
23. Robert Gifford, "The Dragons of Inaction: Psychological Barriers That Limit Climate Change Mitigation and Adaptation," *American Psychologist* 66 (2011): 290–302.
24. Leaf Van Boven, Phillip J. Ehret, and David K. Sherman, "Psychological Barriers to Bipartisan Public Support for Climate Policy," *Perspectives on Psychological Science* 13 (2018): 492–507.
25. Robert Gifford illustrates the different kinds of risk in acting by considering the purchase and use of a PHEV (plug-in hybrid vehicle) as a way of cutting one's carbon footprint. Such a decision carries each of these risks. The PHEV, as new technology, may have problems preventing it from performing as expected (functional); it may be a less safe vehicle than the old SUV traded in for it (physical); it will certainly cost more to purchase (financial); owning it may invite scorn or ridicule from erstwhile friends (social); this may lead to depression (psychological); and the time taken in researching it, deciding to purchase it, learning how to operate it, and perhaps obtaining professional psychiatric help (temporal) may not have been worth it. Who knew that purchasing a PHEV could pose so many kinds of risk? I

did it anyway! Now that I've had a PHEV for two years, I've found all these risks miniscule, so maybe the PHEV option has become less risky since 2011, when Gifford's article was published. Or maybe I am still justifying to myself that I made the right decision?

11 We've Left the Holocene

1. These decisions were ratified by formal votes and reported online on May 21, 2019. See http://quaternary.stratigraphy.org/working-groups/anthropocene/.
2. Life has affected Earth's history before, most notably 3.5 billion years ago, when the cyanobacteria or blue-green algae evolved and began to photosynthesize, releasing oxygen into the atmosphere as a by-product. They made it possible for air-breathing animals like us to evolve, but at no time until now has a single species of life had such an impact as humanity. I'm not sure we should be proud of this achievement.
3. Many individuals may be more easily motivated by a smaller task, one they can get their heads and hearts around. Campaigns to save coral reefs can be effective in changing behavior, just as can campaigns to save polar bears (which also suffer as the world warms), but such campaigns should be presented as important parts of a global effort.
4. It's common to set a global goal of +1.5°C maximum warming, or the more reef-centric goal of putting 30 percent of coastal waters in no-fishing reserves, and then work hard with other partners to reach common ground, only to finally agree that +3.0°C warming or 10 percent of coastal waters in reserves are commendable, mutually satisfactory achievements. This is what usually happens when negotiating large, multinational treaties, as each participant group looks after its own self-interests. We continually forget that Nature does not compromise. If we want a quasi-Holocene world, we need to make the changes that are needed, not reach a compromise that moves us vaguely in the right direction.
5. There is extensive information concerning Hōkūle'a and Polynesian wayfinding at the website of the Polynesian Voyaging Society, http://www.hokulea.com/.
6. As one of the last traditional navigators on 1.3-square-kilometer Satawal, Mau Piailug aided trade and communication with other nearby islands of the Caroline Group that lay far to the west of Hawai'i. He had never sailed Polynesian seas or been south of the equator, but he did have mastery of many Micronesian navigation techniques.
7. Leaf Van Boven, Phillip J. Ehret, and David K. Sherman, "Psychological Barriers to Bipartisan Public Support for Climate Policy," *Perspectives on Psychological Science* 13 (2018): 492–507.
8. In central Ontario, where I live, there exist solid data showing a three-week increase in the duration of the open-water season on our lakes since the mid-1970s. Huaxia

Yao et al., "The Interplay of Local and Regional Factors in Generating Temporal Changes in the Ice Phenology of Dickie Lake, South-Central Ontario, Canada," *Inland Waters* 3 (2013): 1–14. That increase is primarily due to a significantly later average date of ice formation in fall; ice also appears to leave the lakes slightly earlier in spring. The longer open-water season has substantial knock-on effects in summer water temperature, evaporation, nutrient depletion in late summer, and biological productivity of our lakes. There are also consistent trends (becoming earlier) in spring arrival dates for some migratory birds and emergence dates for some insects. Each of these multidecadal trends is best interpreted as a direct consequence of climate change in this region. However, when I tell people about the longer open-water season, it evokes little response. It's *only* three weeks! By contrast, spring flooding appears to be becoming more severe in the sense that hundred-year floods seem to be arriving here every five or six years and our winters have been noticeably more erratic in temperature from week to week the past couple of years. Neither of these apparent changes can yet be firmly attributed to climate change, but I see many of my neighbors citing such weather events as evidence of climate change.

9. Rachel Carson, *Silent Spring* (New York: Houghton Mifflin, 1962). The Flit gun was a manually pumped sprayer for household use named after an insecticidal spray manufactured by Esso and containing 5 percent DDT.
10. I've used information from *DDT: A Review of Scientific and Economic Aspects of the Decision to Ban Its Use as a Pesticide*, prepared for the Committee on Appropriations of the U.S. House of Representatives by EPA, July 1975, EPA-540/1-75-022, available at https://archive.epa.gov/epa/aboutepa/ddt-regulatory-history-brief-survey-1975.html, and from Environmental Defense Fund, https://www.edf.org/about/our-history, for the U.S. portion of the story. The Stockholm Convention now includes 182 members (parties), including 152 signatory countries. Only 6 countries have yet to ratify it (to ratify is to formally commit to abide by its regulations). Stockholm Convention, http://www.pops.int/Home/tabid/2121/Default.aspx.
11. The aurora trout, a color variant of the widespread brook trout (*Salvelinus fontinalis*), was restricted to two small lakes in the Temagami region. It was considered at that time to be either a subspecies of the brook trout or a distinct species.
12. Studying lakes in a region closer to Sudbury's smelters, and well west of Temagami, Richard Beamish and Harold Harvey first drew attention to acidification of lakes in Canada and pointed to industrial air pollution as the likely cause. Richard J. Beamish and Harold H. Harvey, "Acidification of the La Cloche Mountain Lakes, Ontario, and Resulting Fish Mortalities," *Journal of the Fisheries Research Board of Canada* 29 (1972): 1131–1143.
13. Further information about the U.S.–Canada Air Quality Agreement is at Canada–United States Air Quality Agreement: Overview, https://www.canada.ca/en

/environment-climate-change/services/air-pollution/issues/transboundary/canada-united-states-air-quality-agreement-overview.html; and in National Academy of Sciences, *From Research to Reward: The Acid Rain Economy: How the Free Market Tackled an Environmental Challenge* (Washington, D.C.: National Academies Press, 2016), https://doi.org/10.17226/23671, also available online at https://www.nap.edu/read/23671/. Canada and the United States are also parties to the multinational Convention on Long-Range Transboundary Air Pollution, administered by the U.N. Economic Commission for Europe, with fifty-one parties, including all of Europe and the Russian Federation. This convention has used similar approaches to reduce the incidence of acid rain and other noxious air pollution across its member nations.

14. A Dobson unit is a peculiar measurement unit for quantifying trace gases in the atmosphere. It measures the amount of a trace gas in a vertical column through the atmosphere as if each gas were separated out and present as a column of pure gas at the Earth's surface. One DU = such a column 0.001 millimeter tall. The Antarctic changes are in J. C. Farman, B. G. Gardiner, and J. D. Shanklin, "Large Losses of Total Ozone in Antarctica Reveal Seasonal ClO_x/NO_x Interaction," *Nature* 315 (1985): 207–210.
15. In 1974, Mario Molina reported in *Nature* that CFCs in the atmosphere would be rapidly broken down by ultraviolet light, releasing their constituent, highly reactive chlorine and fluorine atoms. These would attack ozone. Mario J. Molina and F. S. Rowland, "Stratospheric Sink for Chlorofluoromethanes: Chlorine Atom-Catalysed Destruction of Ozone," *Nature* 249 (1974): 810–812.
16. Information on the history of the Vienna Convention for the Protection of the Ozone Layer and the Montreal Protocol on Substances That Deplete the Ozone Layer is available in a short online article for the United Nations: Edith Brown Weiss, *The Vienna Convention for the Protection of the Ozone Layer and the Montreal Protocol on Substances That Deplete the Ozone Layer* (2009), available at http://legal.un.org/avl/ha/vcpol/vcpol.html; in Ian Rae, "Saving the Ozone Layer: Why the Montreal Protocol Worked," *The Conversation*, September 9, 2012, https://theconversation.com/saving-the-ozone-layer-why-the-montreal-protocol-worked-9249; and in "Ozone and You: All About Ozone and the Ozone Layer," *UN Environment Programme*, https://ozone.unep.org/ozone-and-you#the-solution-. Recent improvements in the ozone layer, based on information from UNEP's Ozone Secretariat, are reported in Seth Borenstein, "UN Says Earth's Ozone Layer Is Gradually Healing," AP News, November 5, 2018, https://apnews.com/835094e7af61414981259ed69dbb185e.
17. While Richard Beamish was collecting evidence of serious acidification of lakes 60 kilometers away, INCO (International Nickel Co., now Vale) was building the tallest smokestack in Canada—the 381-meter-high "superstack"—to disperse

pollutants from its smelters more widely, thereby reducing impacts in Sudbury itself! The $25-million superstack was completed in 1972. INCO and other smelter and power plant operators were in no hurry to invest in the scrubbing technologies to remove SO_2 and NO_x. In 2017, Vale announced that the superstack would be decommissioned by 2020, replaced by two smaller stacks. This is now under way. Scrubbing technology introduced since 1972 has reduced SO_2 emissions by >88 percent.

18. International treaties always involve politics, and countries (and their leaders) need to learn how to do political negotiations. Former prime minister Brian Mulroney of Canada reminisced about the decade of negotiations on acid rain in Brian Mulroney, "Acid Rain: A Case Study in Canada-US Relations," *Policy Options*, April 1, 2012, http://policyoptions.irpp.org/fr/magazines/harpers-foreign-policy/acid-rain-a-case-study-in-canada-us-relations/.
19. Ironically, these newer chemicals, hydrofluorocarbons, are potent greenhouse gases, so the Montreal Protocol is now moving the world toward their replacement by yet more effective, less damaging refrigerants.

12 Coral Reefs in the Anthropocene

1. I have several purposes in telling these tales of my early underwater experiences in Hawai'i. First, nobody needs scuba certification before they commence studies in marine biology. Second, the increasingly restrictive rules and regulations around use of scuba in academic institutions, driven by our collective enthusiasm for litigation, mean that no university professor would do what Bill Gosline did that day. Students are the poorer for that. Third, I was a remarkably naive young man. And fourth, diving over a coral reef provides a sense of freedom, lightness, and otherworldliness that diving in cold, dark water can never match.
2. Terry P. Hughes et al., "Coral Reefs in the Anthropocene," *Nature* 546 (2017): 82–90; Gareth J. Williams and Nicholas A. J. Graham, "Rethinking Coral Reef Functional Futures," *Functional Ecology* 33 (2019): 942–947; and the seven associated papers in the Special Feature in *Functional Ecology* 33 (2019), spanning pages 948–1034, provide entry to this discussion of the new science needed if we are to steer coral reefs successfully through the Anthropocene.
3. David R. Bellwood et al., "The Meaning of the Term 'Function' in Ecology: A Coral Reef Perspective," *Functional Ecology* 33 (2019): 948–961.
4. Gareth J. Williams et al., "Coral Reef Ecology in the Anthropocene," *Functional Ecology* 33 (2019): 1014–1022.
5. The ethical implications of deliberately assisting ecosystems (or species) to adapt to climate-induced change in conditions must also be considered carefully. Karen

Filbee-Dexter and Anna Smajdor, “Ethics of Assisted Evolution in Marine Conservation,” *Frontiers in Marine Science* 6 (2019): 20, doi: 10.3389/fmars.2019.00020.

6. Madeleine J. van Oppen et al., “Shifting Paradigms in Restoration of the World’s Coral Reefs,” *Global Change Biology* 23 (2017): 3437–3448.
7. An active discussion is now under way within the faith community, reexamining the meaning of dominion and fostering the idea of a moral obligation to act to sustain the natural world. Pope Francis’s encyclical on climate and justice is one fruit of this discussion: Francis, *Laudato Sí: On Care for Our Common Home*, encyclical letter (The Holy See, May 24, 2015), http://w2.vatican.va/content/dam/francesco/pdf/encyclicals/documents/papa-francesco_20150524_enciclica-laudato-si_en.pdf.
8. The Universal Declaration of Human Rights is a milestone document in the history of human rights. The declaration was proclaimed by the U.N. General Assembly in Paris on December 10, 1948, and is available at https://www.un.org/en/universal-declaration-human-rights/.
9. The United Nations Declaration on the Rights of Indigenous Peoples is a resolution adopted by the U.N. General Assembly on September 13, 2007, and is available at https://www.un.org/development/desa/indigenouspeoples/wp-content/uploads/sites/19/2018/11/UNDRIP_E_web.pdf. Ultimately supported by 150 states, with 9 abstaining, it has gained further support in subsequent years. Although nonbinding, it is seen as the most comprehensive international instrument on the rights of Indigenous peoples. Its articles 10, 25–30, and 32 have particular relevance to the topic of land tenure.
10. Eric Dannenmaier, “Beyond Indigenous Property Rights: Exploring the Emergence of a Distinctive Connection Doctrine,” *Washington University Law Review* 86 (2008): 53–110, accessed from *Aboriginal Policy Research Consortium International* (APRCi) 218, https://ir.lib.uwo.ca/aprci/218.
11. By Western-style I mean to include Britain and Europe, their former colonies around the world, and other nations such as Japan and China that have adopted capital-driven, market economies, despite having legal systems that evolved separately from those of western Europe.
12. Aldo Leopold referred to a “land ethic,” the expression of the responsibilities to the land by the owner. An inverse attitude recognizes owners who “extract full value from their land” as somehow better than those who leave it alone to do its own thing, since such owners are contributing to building the economy.
13. Most law governing fisheries, for example, requires that managers ensure that fishing effort not be permitted to exceed that which will provide for maximum sustained yield (MSY) from the fishery stock.
14. I recognize the motivation that propels the desire to eliminate the keeping of animals in zoos, but I reject the suggestion that well-managed zoos cannot keep

animals in a way that is appropriate to the animals' needs. I also know the scientific and conservation value of the research done in zoos and fear the extent of our isolation from the rest of nature that eliminating zoos would bring. On the other hand, the conditions under which most animals were kept in even the best-run circuses are not something any conscious beast should have to endure.

15. Aldo Leopold, "The Land Ethic," in *A Sand County Almanac, with Essays on Conservation from Round River* (Oxford: Oxford University Press, 1966).
16. Christopher D. Stone, "Should Trees Have Standing? Toward Legal Rights for Natural Objects," *Southern California Law Review* 45 (1972): 450–501. The expanded paperback version is Christopher D. Stone, *Should Trees Have Standing? Law, Morality, and the Environment*, 3rd ed. (Oxford: Oxford University Press, 2010).
17. Pachamama is the mother goddess, the principal deity in indigenous belief systems throughout the Andes, and she embodies the Earth, Time, Nature, fertility, fecundity, and the feminine. See "Ecuador Grants Rights to Nature," *Nature Newsblog*, September 29, 2008, http://blogs.nature.com/news/2008/09/ecuador_grants_rights_to_natur.html.
18. Bolivia's Ley de Derechos de la Madre Tierra was passed by the legislative assembly in December 2010. See also John Vidal, "Bolivia Enshrines Natural World's Rights with Equal Status for Mother Earth," *Guardian*, April 10, 2011, https://www.theguardian.com/environment/2011/apr/10/bolivia-enshrines-natural-worlds-rights.
19. Craig M. Kauffman and Pamela L. Martin, "Can Rights of Nature Make Development More Sustainable? Why Some Ecuadorian Lawsuits Succeed and Others Fail," *World Development* 92 (2016): 130–142.
20. The IUCN Red List is the world's most comprehensive inventory of the global conservation status of biological species. Each species on the list has had its risk of extinction critically evaluated by science committees set up under the International Union for Conservation of Nature.

INDEX

Note: Page numbers in *italics* refer to figures.